SpringerBriefs in Molecular Science

Biobased Polymers

Series editor

Patrick Navard, Sophia Antipolis cedex, France

Published under the auspices of EPNOE*Springerbriefs in Biobased polymers covers all aspects of biobased polymer science, from the basis of this field starting from the living species in which they are synthetized (such as genetics, agronomy, plant biology) to the many applications they are used in (such as food, feed, engineering, construction, health, …) through to isolation and characterization, biosynthesis, biodegradation, chemical modifications, physical, chemical, mechanical and structural characterizations or biomimetic applications. All biobased polymers in all application sectors are welcome, either those produced in living species (like polysaccharides, proteins, lignin, …) or those that are rebuilt by chemists as in the case of many bioplastics.

Under the editorship of Patrick Navard and a panel of experts, the series will include contributions from many of the world's most authoritative biobased polymer scientists and professionals. Readers will gain an understanding of how given biobased polymers are made and what they can be used for. They will also be able to widen their knowledge and find new opportunities due to the multidisciplinary contributions.

This series is aimed at advanced undergraduates, academic and industrial researchers and professionals studying or using biobased polymers. Each brief will bear a general introduction enabling any reader to understand its topic.

*EPNOE The European Polysaccharide Network of Excellence (www.epnoe.eu) is a research and education network connecting academic, research institutions and companies focusing on polysaccharides and polysaccharide-related research and business.

More information about this series at http://www.springer.com/series/15056

Yoshiyuki Nishio · Yoshikuni Teramoto
Ryosuke Kusumi · Kazuki Sugimura
Yoshitaka Aranishi

Blends and Graft Copolymers of Cellulosics

Toward the Design and Development
of Advanced Films and Fibers

 Springer

Yoshiyuki Nishio
Division of Forest and Biomaterials Science,
   Graduate School of Agriculture
Kyoto University
Kyoto
Japan

Kazuki Sugimura
Division of Forest and Biomaterials Science,
   Graduate School of Agriculture
Kyoto University
Kyoto
Japan

Yoshikuni Teramoto
Department of Applied Life Science, Faculty
   of Applied Biological Sciences
Gifu University
Gifu
Japan

Yoshitaka Aranishi
Fibers and Textiles Research Laboratories
Toray Industries, Inc.
Mishima, Shizuoka
Japan

Ryosuke Kusumi
Division of Forest and Biomaterials Science,
   Graduate School of Agriculture
Kyoto University
Kyoto
Japan

ISSN 2191-5407   ISSN 2191-5415   (electronic)
SpringerBriefs in Molecular Science
ISSN 2510-3407   ISSN 2510-3415   (electronic)
Biobased Polymers
ISBN 978-3-319-55320-7   ISBN 978-3-319-55321-4   (eBook)
DOI 10.1007/978-3-319-55321-4

Library of Congress Control Number: 2017934614

Printed on acid-free paper

This Springer imprint is published by Springer Nature
The registered company is Springer International Publishing AG
The registered company address is: Gewerbestrasse 11, 6330 Cham, Switzerland

Preface

Today, cellulose and related polysaccharides are well recognized as high-potential polymers to be further materialized for both commodity and specialty uses. The currently more vital research on microscopic composition of cellulosics may be demonstrative of the general recognition. As a part of the compositional research, the present monograph covers basic and applied studies of cellulosic blends and graft copolymers. Polymer blending and grafting techniques can offer opportunities not only to improve the processability and original physical properties of cellulosics, but also to design new, cellulose-core polymeric materials exhibiting wide-ranging or synergistic functions unattainable in gross mechanical mixtures as well as in single-component materials.

The main purpose of this monograph is to survey the fundamental aspects associated with molecular mixing, molecular motions, and supramolecular structuring for cellulosic blends and graft copolymers, and to demonstrate functional aspects linked to their practical applications as advanced films and fibers, as well. Industrially important organic esters of cellulose, such as cellulose acetate, propionate, and butyrate, are employed as representative of the cellulosic component. The monograph is organized into five chapters, each written in a measure commensurate to the respective subject matters as follows: methods for miscibility estimation and structural designing (Chap. 1); typical examples of detailed characterization (Chaps. 2–4); embodiment of high-functional optical films (Chaps. 2 and 4), biodegradable/biocompatible moldings (Chaps. 3 and 4), and melt-spun green fibers (Chap. 5).

The constitutive chapters have their own share to accomplish the above main purpose in reasonable correlation with each other. A sequence of results compiled into this book will provide useful suggestions on the designing of functionality-rich multicomponent materials based on cellulosics, which in turn will contribute toward more expanding the availability of cellulose. Therefore, this book will hopefully be helpful to many scientists and technologists engaged on cellulose and renewable materials research in academia and in industry, and, of course, to graduate students touching bio-based polymers in universities.

Finally, I would like to express my sincere gratitude to Dr. Patrick Navard of Ecole des Mines de Paris, CNRS, France, who is the president of the European Polysaccharide Network of Excellence (EPNOE), for his encouragement and helpful discussions as well as for his kind invitation to contribute to this interesting series of SpringerBriefs. I am also grateful to the Springer staff involved in coordinating this publication, for their kind assistance to the overall editing work.

Kyoto, Japan   Yoshiyuki Nishio
November 2016   On behalf of the authors

Contents

Abbreviations

ACMO	Acryloyl morpholine
Acyl-Ch	Acyl chitin
AFM	Atomic force microscopy
AGU	Anhydroglucose unit
ATR-FTIR	Attenuated total reflection Fourier transform infrared spectroscopy
ATRP	Atom transfer radical polymerization
BL	(R,S)-β-Butyrolactone
CA	Cellulose acetate
CAB	Cellulose acetate butyrate
CAP	Cellulose acetate propionate
CAV	Cellulose acetate valerate
CB	Cellulose butyrate
CBV	Cellulose butyrate valerate
CC	Cellulose caproate
CE	Cellulose ester
CEn	Cellulose enanthate
ChA	Chitin acetate
ChB	Chitin butyrate
ChC	Chitin caproate
ChP	Chitin propionate
ChV	Chitin valerate
CL	ε-Caprolactone
CP	Cellulose propionate
CP-MAS	Cross-polarization and magic-angle spinning
CPV	Cellulose propionate valerate
CV	Cellulose valerate
D	Average crystallite size
DD	Degree of deacetylation
ΔH_m (or ΔH_f)	Heat of fusion

DMA	Dynamic mechanical analysis
DMAc	N,N-Dimethylacetamide
DMF	N,N-Dimethylformamide
Δn	Birefringence
DP_s	Degree of polymerization in the side chain
DRS	Dielectric relaxation spectroscopy
DS	Degree of substitution
DSC	Differential scanning calorimetry
E'	Dynamic storage modulus
E''	Dynamic loss modulus
FE-SEM	Field emission scanning electron microscope
FT-IR	Fourier transform infrared spectroscopy
IPN	Interpenetrating polymer network
K_g	Nucleation factor for folded chain crystallization
LDH	Layered double hydroxide
M	Molecular weight per AGU containing ester and pendant PHA side chains
MMA	Methyl methacrylate
M_n	Number-average molecular weight
MS	Molar substitution
M_{side}	Number-average molecular weight of polyester chains introduced onto CE
N	Number of carbons in the normal acyl substituent
n	Avrami exponent
PACMO	Poly(acryloyl morpholine)
PCL	Poly(ε-caprolactone)
PEG	Poly(ethylene glycol)
PET	Poly(ethylene terephthalate)
PHA	Poly(hydroxy alkanoate)
PHB	Poly(3-hydroxybutyrate)
PHBV	Poly(3-hydroxybutyrate-co-3-hydroxyvalerate)
PLA	Poly(lactic acid)
PLLA	Poly(L-lactic acid)
PMMA	Poly(methyl methacrylate)
POM	Polarized optical microscope
PVA	Poly(vinyl alcohol)
PVAc	Poly(vinyl acetate)
PVAVAc	Poly(vinyl acetate-co-vinyl alcohol)
PVL	Poly(δ-valerolactone)
PVP	Poly(N-vinyl pyrrolidone)
P(VP-co-MMA)	Poly(N-vinyl pyrrolidone-co-methyl methacrylate)
P(VP-co-VAc)	Poly(N-vinyl pyrrolidone-co-vinyl acetate)
PVPh	Poly(vinyl phenol)
$t_{1/2}$	Half-time of crystallization
T_1^H	Proton spin–lattice relaxation time

$T_{1\rho}^{H}$	Proton spin–lattice relaxation time in the rotating frame
T_g	Glass transition temperature
T_{ic}	Isothermal crystallization temperature
T_m^{eq}	Equilibrium melting temperature
VAc	Vinyl acetate
VL	δ-Valerolactone
VP	N-Vinyl pyrrolidone
WAXD	Wide-angle X-ray diffraction
w_i	Weight fraction of component i
X_c	Crystallinity index
β	Non-exponential parameter indicating the degree of distribution of relaxation time
ε''	Imaginary part of a complex dielectric function
ε'	Real part of a complex dielectric function
μ	Viscometric interaction parameter
σ_e	Fold surface free energy per unit area of chain-folded crystal

Chapter 1
General Remarks on Cellulosic Blends and Copolymers

Yoshiyuki Nishio

Abstract This chapter is a general introduction to the present monograph and first describes the significance of the studies on "*Blends and Graft Copolymers of Cellulosics*" in the research field of compositional materials based on cellulose and related polysaccharides. Secondly, some technical key-terms and methods used for characterizing cellulosic blends and graft copolymers are explained. Finally, the outline of this book is provided by summarizing the main subjects of the constituting chapters with a perspective to tie together the subjects.

Keywords Blends · Cellulose · Cellulose esters · Functionality · Graft copolymers · Microcomposition · Miscibility · Multicomponent materials · Synergistic effect

1.1 Introduction: In the Stream of Microcomposition Research

Today, polysaccharides represented by cellulose and its relatives (glucans) are well recognized as sustainable bioresources and high-potential polymers to be further materialized for both commodity and specialty uses. For instance, cellulosic polymers exhibit various characteristics at the molecular and supramolecular levels; e.g., the polymer molecules have side-group reactivity (allowing further derivatization), hydrogen-bonding formability, enzymatic degradability, semi-rigidity, and chirality, and they can form distinct higher-order structures, such as fibrous crystalline entities (conditionally extractable in a nano-sized section), lyo-gels, and even liquid crystals. Only one class of polymers (generically cellulosics) is endowed with so many fascinating characteristics. This drives many researchers to make advanced use of cellulosics in diverse fields involving functional materials, although accurate understanding is still required for the individual characteristics or attributes.

A viable approach to multifaceted applications utilizing the natural bioresource, cellulose, involves designing multifunctional or high-performance materials based on cellulosics via microcomposition at the single-molecule level or nanofibrils with

© The Author(s) 2017
Y. Nishio et al., *Blends and Graft Copolymers of Cellulosics*,
Biobased Polymers, DOI 10.1007/978-3-319-55321-4_1

other chemical ingredients [1–8]. The ingredients commonly used as counter component combined with cellulosics are other organic polymers, but inorganic compounds, such as layered clays, may also be employed as an additional component as a nanoscale filler. However, nanohybridization using various inorganics (e.g., silica, metal oxides, and apatites) is more attractive when cellulose and related polysaccharides are the matrix or template components and are assembled in a particular state such as gel or liquid crystal [1–3, 9–11].

In a traditional concept, we expect a "complementary" effect for the properties of the compositional materials; i.e., a certain deficiency of one component will be partly compensated by the advantage of the other component if a binary system is considered. This is an important universal concept, but an up-to-date material design using a multicomponent strategy will require attainment of widely changeable properties with varying component compositions and even a drastic "synergistic" improvement of some properties that are linked to a new functionality. Figure 1.1 illustrates the ideal cases of such synergism for a material system composed of component I (C-I) and II (C-II). In case 1 (Fig. 1.1a, upper diagram), desirable property **A** varies in intensity with the relative composition of the C-I/C-II mixture and shows the optimum at an intermediate composition. Case 2 (Fig. 1.1a, bottom diagram) is the extreme where property (or capability) **X** (unrealized in C-I and C-II) arises in a definite range of the binary composition. Cases 3 and 4 are illustrated in Fig. 1.1b and represent synergistic improvements to undesirable properties **B** and **Y**, respectively, that should be reduced or suppressed for practical use. All four cases satisfy a large shift of the material properties to a favorable course from the linear additive rule of mixing. In empirical deduction, when the

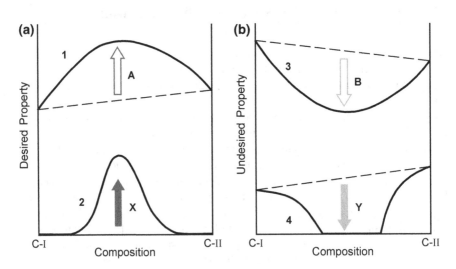

Fig. 1.1 Illustrations of the synergistic behavior in property idealized for a material system composed of components C-I and C-II. Patterns 1 and 2 depicted in part **a** are concerned with desired properties **A** and **X**, respectively. Patterns 3 and 4 in part **b** refer to undesired properties **B** and **Y**, respectively

system is a gross bulk mixture, the possible variation in a usable property would be limited and hardly show such a welcome deviation from the simple additive rule, even though the C-I/C-II composition is changed; actually, the wide compositional change is often difficult for the fabrication process into dimensioned materials. Instead, the more intimate incorporation at a hyperfine structural level (usually $<10^{-8}$ m) of at least one component will increase the chance of discovering that kind of synergism in material properties. This approach can hold good in composing cellulosics as the component, and the research is interesting and significant in view of their ready availability with diverse structural features.

Among the extensive studies on microcompositions of cellulosics, the present monograph focuses on the blends that are miscible or compatible with other polymers and on the graft copolymers, the trunk chain of which is made up of a carbohydrate backbone. The terms of polymer blending and copolymerization are rather trite compared to that for the nanocomposite or nanohybrid, but are basically pivotal and far-reaching for designing polymer-based multicomponent materials. In fact, the knowledge of miscibility, domain formation, interaction, etc., is useful for enhancing the performance of various gross- and nano-composites, for example, by improving the adhesion of ingredient interfaces by chemical modification of the bulk surfaces of the raw materials.

This monograph is not a comprehensive review, and the cellulosic polymers covered herein are mainly organic esters of cellulose (CEs), such as cellulose acetate, propionate, and butyrate. Concerning blends of unmodified cellulose with synthetic polymers, a general scheme for their preparation and characterization has been concisely described in previous reviews [1, 12, 13]. This monographic book will summarize the total recent progress in fundamental characterization and functional development of CE blends and copolymers, although the constitutive chapters center on the authors' research achievements. Synergistic effects will be demonstrated for some properties, including thermal processability, in connection with the practical applications of the cellulose-core materials to new advanced films, membranes, fibers, and so forth.

1.2 Terminology

Several key specialty terms are common to the study of cellulosic blends and copolymers. The definition or usage of these terms in this treatise is clarified below.

1.2.1 Structural Parameters of Cellulose Derivatives

Cellulose is a $\beta(1 \rightarrow 4)$-linked glucan and has three hydroxyl groups in the anhydroglucose unit (AGU) (Fig. 1.2a). Because of the hydroxyl reactivity, a variety of cellulose derivatives can be synthesized, and industrially established grades of the ester/ether derivatives are readily available. In addition to the molecular weight, an important structural parameter that characterizes cellulose derivatives is the *degree of substitution* (DS), which is defined as the average number of substituted hydroxyls per AGU. A derivative of DS = 1.5 is schematically shown in Fig. 1.2b (upper). If repeatable chain growth is possible for the used substituent, then another parameter, MS (*molar substitution*), is defined as the average number of substituent groups introduced per AGU. Figure 1.2b (bottom) illustrates a derivative structure of MS = 4 and DS = 2. Use of MS is a common practice for characterizing, for example, hydroxyalkylether derivatives of cellulose, and similar applicability can extend to cellulosic graft copolymers.

DS and MS can be determined by conventional NMR measurements. An example of the determination is briefly given here. Figure 1.3 shows ^{1}H NMR spectra measured for samples of cellulose butyrate (CB) and CB-*graft*-poly (ε-caprolactone) (CB-*g*-PCL) in CDCl$_3$ solvent. The copolymer was obtained by a ring-opening polymerization of ε-caprolactone initiated at the residual hydroxyl positions of the CB sample (see Chap. 4). In the CB spectrum (Fig. 1.3a), a resonance peak area derived from the methyl protons of the butyryl groups is designated as **A**, and a total peak area from the protons of the glucopyranose unit is

Fig. 1.2 a Structural formula of cellulose; **b** schematic illustrations of a cellulose derivative of DS = 1.5 (*upper*) and that of MS = 4 and DS = 2 (*bottom*)

Fig. 1.3 ¹H NMR spectra of
a CB of butyryl DS = 2.10
and **b** CB-*g*-PCL of butyryl
DS = 2.10 and oxycaproyl
MS = 2.33, measured in
CDCl₃

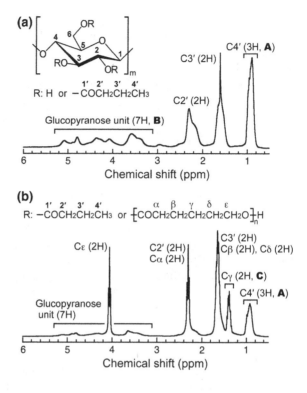

designated as **B**. Then, the butyryl DS (= 2.10) is evaluated by the following
equation:

$$DS = \frac{A/3}{B/7} \qquad (1.1)$$

From the spectrum of CB-*g*-PCL (Fig. 1.3b), the MS value (= 2.33) of this graft
copolymer is determined from the following equation:

$$MS = DS \times \frac{3C}{2A} \qquad (1.2)$$

where **C** is a resonance peak area from the C_γ methylene protons of the PCL
side-chains, and DS refers to the butyryl substituent. Using the MS data and calculated formula weights of the CB repeating unit and PCL-oxycaproyl unit, the
weight fraction of the grafted PCL component, W_{PCL}, can also be determined;
$W_{PCL} = 0.46$ for the example given in Fig. 1.3b.

The *degree of polymerization of the side chains*, DP_s, is an additional important
parameter for the molecular characterization of graft copolymers. The average value
may be simply estimated from MS/DS_{graft}, where DS_{graft} indicates the DS of the
introduced grafts. It is easy to quantify DS_{graft} (and therefore DP_s) from the NMR

data when a ^{1}H NMR signal reflecting the side-chain terminal can be distinguished. This is the case for cellulose acetate-*graft*-poly(L-lactic acid) [14] (see Chap. 4), but not for CB-*g*-PCL. In the latter case, an apparent value of DP_s is calculated by assuming $DS_{graft} = 3 -$ (butyryl DS). Again, using the example shown in Fig. 1.3b, we find a value of 2.59 for the apparent DP_s.

1.2.2 Miscibility of Polymer Blends

In thermodynamics, *miscibility* refers to the single-phase at a molecular level for a system usually consisting of two chemical compounds. By way of induction, "miscible polymer blend" would imply that the mixing state of the constituents is homogeneous at a segmental level of the macromolecular chain. Correctly, the term "*miscible*" only suggests that the level of molecular mixing is adequate to yield the macroscopic properties expected of a single-phase material [15]. For practical purposes, the classification of polymer blends as miscible is made by detecting a single glass transition whose temperature (T_g) is usually intermediate between those of the component polymers. This criterion is also taken in the present treatise (see Chaps. 2 and 3). However, the miscibility assessment depends on the type and condition of the test applied in the study [16].

Compatibility has often been used synonymously with miscibility [17]. Nowadays, compatibility seems to be a more utilitarian term indicating a commercially attractive polymer mixture, e.g., the kind that is normally homogeneous to the eye (in the film state) or frequently shows an enhanced mechanical property over the constituent polymers [16].

Generally, a certain *specific interaction* between the two constituent macromolecules is necessary to achieve miscible binary polymer blends. The reason is instantly explicable in terms of the thermodynamic condition of ideal mixing, i.e., the free energy of mixing, ΔG_{mix} (= $\Delta H_{mix} - T \Delta S_{mix}$), must be negative. For high molecular weight polymers, the entropy of mixing, ΔS_{mix}, is reasonably assumed to be negligible. Therefore, the enthalpy or heat of mixing, ΔH_{mix}, must be negative for miscibility to occur, indicating the necessity of an exothermic interaction. Hydrogen bonding serves as the relevant interaction. Dipolar or ionic interactions can also provide similar properties. The importance of such specific interactions defined by the proton or electron donor–acceptor concept has been well recognized in many examples of miscible polymer blends [15–18].

Instead of the specific interactions described above, another type of interaction is often the necessary driving force for miscibility. It is the "*intramolecular repulsion*" of at least one component in the binary polymer blend. An explicit example may be found when the component is a copolymer in which two monomer species of mutually repellent characters are covalently linked in a random fashion (see Chap. 2). If the intramolecular repulsion is strong, the copolymer can intimately mix with the other component rather than aggregate among the same copolymer chains.

1.3   Methods for Miscibility Estimation

The experimental methods used to assess polymer–polymer miscibility are summarized below and are roughly divided into two groups: thermal transition analysis and spectroscopy. Microscopy is not included here; however, needless to say, the application is more powerful for examining the bulk-surface or internal morphology of polymer blends and copolymers.

1.3.1 T_g Measurements

As stated in the preceding section, the well-established method for determining polymer–polymer miscibility is the detection of the glass transition of the blend versus those of the unblended constituents. Differential scanning calorimetry (DSC) and dynamic mechanical analysis (DMA) are widely used for such measurements. In DSC thermograms, the glass transition of the sample is signaled by a baseline shift of the heat flow. This shift is equivalent to the discontinuity in specific heat arising when the temperature of the heated sample passes through the transition range. In most cases, the glass transition temperature T_g is taken as the onset point or midpoint of the baseline shift. Generalized DSC data are depicted in Fig. 1.4a; a miscible polymer blend (A/B = 50/50) shows a single T_g intermediate between the

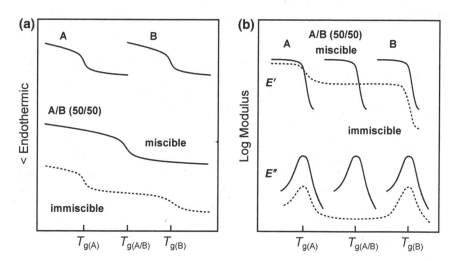

Fig. 1.4 Schematic representation of the glass transition behavior as observed by **a** DSC and **b** DMA measurements for a miscible polymer blend (A/B = 50/50), in comparison with the behaviors of unblended constituents A and B. In counterpoint, the case expected for an immiscible (two-phase) blend is illustrated using a *dotted line-curve*

two T_gs of the constituent polymers (A and B). For an immiscible blend, we should observe the two separate T_gs, as illustrated by the dotted line.

In DMA, a polymer material is subjected to a small-amplitude cyclic tensile (or shear) deformation, and its viscoelastic response provides information on the glass transition and other sub-transitions. The dynamic storage modulus E' (or G' for shear) and loss modulus E'' (or G'' for shear), and the loss tangent (tan δ) are obtained as a function of temperature at a nominal frequency, typically 0.5–100 Hz. The glass transition appears as a so-called primary dispersion signal corresponding to the α relaxation of an amorphous material, reflecting the segmental micro-Brownian motions of the polymer main-chain. Figure 1.4b illustrates ideal E' and E'' data associated with the principal relaxation for a miscible polymer blend (again, A/B = 50/50), indicating a single glass transition situated between those of unblended A and B. The T_g of the blend is defined as the temperature at which tan δ or E'' assumes a maximum value within the transition range. Strictly, the use of the peak position of E'' may be preferable. The loss factor, tan δ (= E''/E'), is formally a ratio of the two dynamic moduli, both rapidly changeable with temperature in the transition range; the maximum position as T_g is sometimes indiscernible.

In estimation of the miscibility for a pair of polymers, it is important to make a plot of T_g versus blend composition. Observation of a single, composition-dependent T_g over the whole composition range is the right sign of total miscibility, and the presence of two T_g values indicates immiscibility or partial miscibility. DSC excels in facility of the collection of T_g data. In this instrumentation, accurate control of the thermal history of the used sample (typically 5–20 mg) can be maintained by programmed heating and cooling cycles; additionally, analysis of the melting and crystallization behavior is feasible for polymer blends showing crystallinity. With DMA, only films or filamentous objects made from polymer mixtures are usable in the T_g measurement, and attention should be paid to the treatment history of the test sample. An advantage of DMA is that we can discuss the effect of blending on the low-temperature secondary relaxations associated with local-part motions of the component polymers. Besides this, DMA can form a general estimate of the thermo-mechanical performance of the objective material.

In many studies of polymer blends, we find good qualitative agreement in the T_g–composition dependence between the DSC and DMA results. However, there is a case of conflict; e.g., a particular blend may be judged miscible by DSC but heterogeneous by DMA. This fact indicates that, regarding T_g detection, the two techniques are responsive to similar molecular relaxations occurring over different region sizes [19]. According to a generally accepted opinion [15, 16, 19, 20], the level of molecular mixing to yield a single T_g in DMA for polymer blends may be ~ 15 nm as an upper limit of the possible domain size. In DSC, the limit would increase to a certain larger value, e.g., ~ 25 nm.

1.3.2 Spectroscopic Measurements

In the primary case of miscibility achievement where specific interactions are involved (see Sect. 1.2.2), the two component polymers should possess mutually different interacting groups within their macromolecular chains. Infrared (IR) spectroscopy, usually with a Fourier transform (FT) interferometer, can be a powerful tool to substantiate the interactions. FT-IR measurements have revealed interactions of hydrogen-bonding type for many miscible blends by detecting frequency shifts of the specific IR bands of functional groups belonging to the component polymers [18]. In CE blends (see Chap. 2), hydroxyl (O-H) and carbonyl (C=O) groups often participate in the hydrogen-bonding interaction; the changing shape and wavenumber position of the stretching vibrational bands are followed as a function of the blend composition [21, 22].

Solid-state ^{13}C NMR spectroscopy with cross polarization and magic angle spinning (CP-MAS) equipment is also useful for characterizing the miscibility of polymer blends. The interaction of the blend components through hydrogen bonding can cause changes in line shape and/or frequency of the ^{13}C resonance peaks in the CP-MAS NMR spectra of the blends compared to those of the unblended components. With regard to cellulosic blends, this has been clearly exemplified for highly miscible systems [22–24].

Another important use of solid-state NMR for polymer blends is to approximate the size of heterogeneity, L, which indicates the miscibility level [25–27]. This is commonly conducted by quantifying T_1^H and $T_{1\rho}^H$, which are the so-called proton spin-lattice relaxation time and that in a rotating frame, respectively. Each relaxation process is governed predominantly by ^1H-spin diffusion that is mediated by ^1H–^1H dipolar coupling. Practically, these relaxation times are determined by analysis of the decaying carbon resonance signals detected in the respective relevant CP-MAS NMR experiments. Commonly observed $T_{1\rho}^H$ values are on the order of 10^{-3}–10^{-2} s, while T_1^H is much longer, typically ranging from ~ 1 to 5 s. For a binary blend, if equalization in $T_{1\rho}^H$ between the constituent polymers is observed, the system is homogeneous within a few nanometers (e.g., $L \leq$ 2–4 nm). If far separate values of T_1^H are obtained, the averaged domain size is taken to be larger than several tens of nanometers (e.g., $L > 40$ nm), to give an extreme example. These limits of L are evaluated by approximation to an effective path length of the ^1H-spin diffusion permitted over a period of time, $t = T_{1\rho}^H$ or T_1^H, from the following equation [25]:

$$L \cong (6Dt)^{1/2} \qquad (1.3)$$

where D is the diffusion coefficient, which is usually assumed to be $\sim 10^{-16}$ m^2/s in organic polymer materials. In this treatise, the spin-diffusion measurements will be very helpful in detailing the intercomponent mixing state in films of CE-based blends (Chaps. 2 and 3) and graft copolymers (Chap. 4).

Figure 1.5 shows construction of a type of "ruler" to estimate the homogeneity scale in multicomponent polymer systems (here, blends and copolymers). The

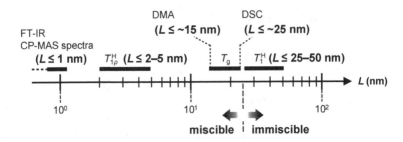

Fig. 1.5 Standard ruler to estimate the homogeneity scale in multicomponent polymer systems. The approximate limits of heterogeneity examined by various techniques are marked on a logarithmic coordinate axis of the domain size L

typical limits of heterogeneity detected by various techniques are marked on a logarithmic coordinate axis of the domain size L (in nm). The T_g measurement is sensitive to heterogeneities with sizes larger than ~ 15 nm (DMA) or ~ 25 nm (DSC), as previously noted. In the solid-state CP-MAS NMR and FT-IR spectra measurements, the heterogeneity of mixing is assumed to be monitored on a scale of ~ 1 nm (in view of the main cause of the frequency shifts, i.e., the interaction between chemical moieties of a few angstroms). The analysis of $T_{1\rho}^H$ and T_1^H is complementary to the CP-MAS/FT-IR spectra and T_g measurements; the relaxation times are useful for estimating the upper limit of heterogeneity that lies commonly in a range of approximately 2–5 nm ($T_{1\rho}^H$) or 25–50 nm (T_1^H). Thus we can ensure the approximate scale of homogeneous mixing in a target system by putting several techniques to good use for comparison.

1.4 Supplementary Techniques

This section provides descriptions of two techniques: dielectric relaxation and fluorescence polarization measurements. These are used to feature the molecular dynamics and orientation behavior, respectively, in CE blends and graft copolymers. Analysis of molecular motions is of great significance in understanding of the thermoplasticization of cellulose by polymer grafting. Molecular orientation and optical anisotropy, induced generally by deformation of polymer films, are important in relation to the applicability of the CE-based materials to various optical media.

1.4.1 Dielectric Relaxation Spectroscopy

Dielectric relaxation spectroscopy (DRS) is a useful tool for investigating molecular dynamics of polymer materials in a widely extended time scale [28] if the moving site of the polymer main-chain and/or attached side-chains has a permanent dipole moment. A small fluctuation of polar units (e.g., -C-O-O-C-) can be detected in dielectric relaxation spectra, whereas it is hard to detect such a small motion by DMA and NMR techniques.

Dielectric relaxations are generally described as a combination of the real (ε') and imaginary (ε'') parts of a complex dielectric function. The relaxation processes are each detected as a discrete dispersion signal and can be simulated using the following Havriliak-Negami equation [29]:

$$\varepsilon^* = \varepsilon_\infty + (\varepsilon_s - \varepsilon_\infty)/\left\{1 + (i\omega\tau)^{\beta_i}\right\}^{\alpha_i} \tag{1.4}$$

where ε^* is the complex permittivity, ε_∞ and ε_s denote the limits of ε' to higher and lower frequencies, ω is the angular frequency of the measurement ($\omega = 2\pi f$ with normal frequency f), τ is the dielectric relaxation time, and α_i and β_i are parameters that characterize the shape of the relaxation time distribution ($0 < \beta_i \leq 1$, $0 < \alpha_i\beta_i \leq 1$). If α_i is normalized to 1, Eq. (1.4) is reduced to a Cole-Cole relationship [30], in which β_i indicates the degree of distribution of the relaxation time associated with a dynamic process. A situation of $\alpha_i = \beta_i = 1$ leads to the simplest Debye function considering no distribution of the relaxation time. Figure 1.6 illustrates frequency dependences of ε' and ε'' simulated in terms of these two types of functions.

If a dispersion signal partly overlaps with an ascent of direct current (dc) conductivity (this is often observed for the primary α relaxation), the following equation including a correction term ($-i(\sigma_{dc}/\omega\varepsilon_0)$) should be adopted to extract the net relaxation process:

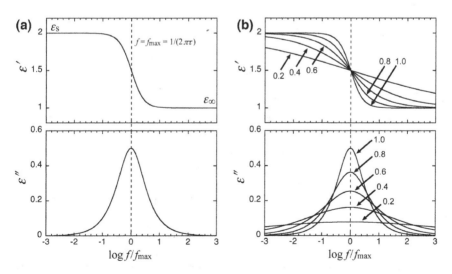

Fig. 1.6 Dielectric relaxation curves that follow the Havriliak-Negami equation (Eq. 1.4) with **a** $\alpha_i = \beta_i = 1$ (Debye type) and **b** different β_is under $\alpha_i = 1$ (Cole-Cole type)

$$\varepsilon^* = \varepsilon_\infty + (\varepsilon_s - \varepsilon_\infty)/\left\{1 + (i\omega\tau)^{\beta_i}\right\}^{\alpha_i} - i(\sigma_{dc}/\omega\varepsilon_0) \qquad (1.5)$$

where σ_{dc} and ε_0 are the dc conductivity and the permittivity of a vacuum, respectively. In the DRS study conducted for a series of CE-*graft*-aliphatic polyesters (Chap. 4), the major quantities of the dielectric relaxation, ε'', τ, and β_i, are discussed through determination using Eqs. (1.4) or (1.5) with $\alpha_i = 1$.

1.4.2 Fluorescence Polarization Measurement

The fluorescence polarization technique for estimating the molecular orientation in polymer solids is well established [31, 32]. Usually the sample to be deformed (here, a uniaxially drawn film) contains a slight amount of a fluorescent probe, e.g., a stilbene derivative with a molecular axis, M, of ~ 2.5 nm (see Fig. 1.7a), which should be dispersed in the amorphous regions of the polymer material. In the optical system shown in Fig. 1.7b, the orientation of M is specified by a set of polar and azimuthal angles (ω, φ) in a sample coordinate frame O-XYZ, where the Z-axis is aligned in the draw direction of the film sample. The molecular probe is excited by polarized light (wavelength, λ_{ex}) through a polarizer (P_1) with transmission axis P_1, and the subsequently emitted fluorescence light (wavelength $\lambda_f > \lambda_{ex}$) is detected as a polarized component through an analyzer (P_2) with transmission axis P_2. The

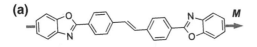

4,4'-bis(2-benzoxazolyl)stilbene (BBS)

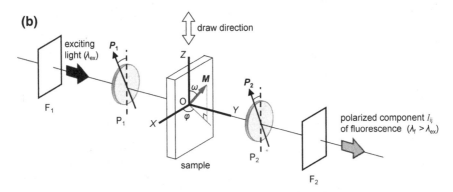

Fig. 1.7 Standard setup of fluorescence polarization measurements: **a** illustration of a fluorescent probe with a molecular axis M to be incorporated into the sample (O-XYZ in part b); **b** optical system for measuring the polarized component of fluorescence intensity (I_{ij}), equipped with a kit of monochromatic filter (F_1), polarizer (P_1,) analyzer (P_2), and cut-off filter (F_2)

probability of excitation and that of detection are related, virtually, to the square of a scalar product $(M \cdot P_1)$ and that of $(M \cdot P_2)$, respectively. Because of the two-fold selectivity, the overall intensity (I_{ij}) of the polarized component of fluorescence obtained from the system is a function of the second and fourth moments $(<\cos^2 \omega>$ and $<\cos^4 \omega>)$ of molecular orientation, which is defined as follows:

$$<\cos^k \omega> = \int_0^{2\pi} \int_0^{\pi} \cos^k \omega N(\omega, \varphi) \sin \omega \, d\omega \, d\varphi \, (k = 2, 4) \qquad (1.6)$$

where $N(\omega, \varphi)$ is a normalized function representing the molecular orientation distribution. Conventionally, these moments can be evaluated from intensity measurements of four polarized components, I_{ZZ}, I_{ZX}, I_{XX}, and I_{XZ} [32]; I_{ZZ} refers to a component observed when the two axes, P_1 and P_2, are both set parallel to the Z-axis, and I_{ZX} is of a component observed when P_1 and P_2 are parallel to the Z- and X-axes, respectively. I_{XX}, and I_{XZ} are defined in a similar way.

The fluorescence technique is employed to obtain information about the overall degree and type of molecular orientation in drawn films of CE blends (Chap. 2) and graft copolymers (Chap. 4). In this connection, birefringence quantification is also performed for the drawn films to estimate the state of optical anisotropy induced therein.

1.5 Outline of the Monograph

This monograph is organized into five chapters. Altogether, the main purpose is to survey the fundamental aspects associated with molecular mixing, molecular motions, and possible supramolecular structuring (e.g., crystallization) for cellulosic blends and graft copolymers, and to demonstrate functional aspects linked to their practical applications as advanced films and fibers. Industrially established CEs, namely, cellulose acetate (CA), cellulose propionate (CP), and cellulose butyrate (CB), are employed to represent the cellulosic component. The five chapters each accomplish the above purpose in cooperation with each other.

This chapter describes the general background and common technical terms and methods for studying cellulosic blends and graft copolymers and provides the outline of this monograph (see Fig. 1.8).

In Chap. 2, the blend miscibility of CA, CP, and CB with vinyl copolymers (mainly comprising the N-vinyl pyrrolidone (VP) unit) is characterized as a function of the acyl DS of the respective CE components and the copolymer composition of the counter component. Inter- or intra-molecular interactions that contribute to the miscible pairings are clarified. Applications of the miscible blends to functional optical films (e.g., film elements in the displays) and permeation-selective membranes are suggested.

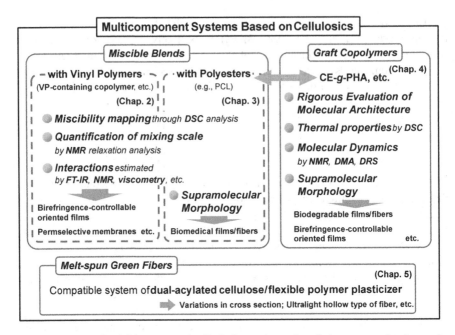

Fig. 1.8 Framework of this monograph, displaying major topics of chapters correlated to each other

In Chap. 3, a typical biodegradable polyester, poly(ε-caprolactone) (PCL), is selected for the counter component of the CE blends. In addition to CA, CP, and CB, a few CE samples are prepared with acyl substituents larger in the alkyl-carbon number N than butyryl one ($N = 4$), and at different DSs. The miscibility and melt-crystallization behaviors are investigated for blends of PCL with various CEs and systematized in terms of N and DS of the CEs. A comparable study is performed for a series of acylated chitin/PCL blends, and their applicability to biomedical materials is highlighted.

In Chap. 4, manifold structure–property relationships are discussed for CE-*graft*-aliphatic polyesters (or poly(hydroxyalkanoate)s), CE-*g*-PHAs, which are designed as environmentally conformable and thermally moldable materials. Particular attention is turned to molecular dynamics of the CE trunk and PHA graft chains, and to the supramolecular morphology developed via heat treatment and/or enzymatic hydrolysis of the molded copolymer films. The chapter deals with another graft copolymer synthesized by atom transfer radical polymerization; its oriented films are evaluated for optical functionality.

In Chap. 5, successful production of a melt-spun cellulosic fiber is exemplified. We show that the following two chemical techniques are important for the melt-spinning process at reasonable temperatures: (1) adequate modification of cellulose, e.g., into dual-acylated cellulose exhibiting a modest thermoplasticity and

(2) miscible blending of the modified cellulose with a highly flexible polymer as a viscosity controller.

A sequence of results compiled into this book will provide useful suggestions for designing functionality-rich multicomponent materials based on cellulosics.

References

1. Nishio Y (2006) Material functionalization of cellulose and related polysaccharides via diverse microcompositions. Adv Polym Sci 205:97–151. doi:10.1007/12_095
2. Habibi Y, Lucia LA, Rojas OJ (2010) Cellulose nanocrystals: chemistry, self-assembly, and applications. Chem Rev 110:3479–3500. doi:10.1021/cr900339w
3. Klemm D, Kramer F, Moritz S, Lindström T, Ankerfors M, Gray DG, Dorris A (2011) Nanocelluloses: a new family of nature-based materials. Angew Chem Int Ed 50:5438–5466. doi:10.1002/anie.201001273
4. Moon RJ, Martini A, Nairn J, Simonsen J, Youngblood J (2011) Cellulose nanomaterials review: structure, properties and nanocomposites. Chem Soc Rev 40:3941–3994. doi:10.1039/c0cs00108b
5. Eichhorn SJ (2011) Cellulose nanowhiskers: promising materials for advanced applications. Soft Matter 7:303–315. doi:10.1039/c0sm00142b
6. Lin N, Huang J, Dufresne A (2012) Preparation, properties and applications of polysaccharide nanocrystals in advanced functional nanomaterials: a review. Nanoscale 4:3274–3294. doi:10.1039/c2nr30260h
7. Carlmark A, Larsson E, Malmström E (2012) Grafting of cellulose by ring-opening polymerization—a review. Eur Polym J 48:1646–1659. doi:10.1016/j.eurpolymj.2012.06.013
8. Teramoto Y (2015) Functional thermoplastic materials from derivatives of cellulose and related structural polysaccharides. Molecules 20:5487–5527. doi:10.3390/molecules 20045487
9. Giese M, Blusch LK, Khan MK, MacLachlan MJ (2015) Functional materials from cellulose-derived liquid-crystal templates. Angew Chem Int Ed 54:2888–2910. doi:10.1002/anie.201407141
10. Nishio Y, Sato J, Sugimura K (2016) Liquid crystals of cellulosics: fascinating ordered structures for the design of functional material systems. Adv Polym Sci 271:241–286. doi:10.1007/12_2015_308
11. Hamad WY (2016) Photonic and semiconductor materials based on cellulose nanocrystals. Adv Polym Sci 271:287–328. doi:10.1007/12_2015_323
12. Nishio Y (1994) Hyperfine composites of cellulose with synthetic polymers, Chap. 5. In: Gilbert RD (ed) Cellulosic polymers, blends and composites. Carl Hanser, Munich
13. Vigo TL (1998) Interaction of cellulose with other polymers: retrospective and prospective. Polym Adv Technol 9:539–548. doi:10.1002/(SICI)1099-1581(199809)9:9<539:AID-PAT813>3.0.CO;2-I
14. Teramoto Y, Nishio Y (2003) Cellulose diacetate-*graft*-poly(lactic acid)s: synthesis of wide-ranging compositions and their thermal and mechanical properties. Polymer 44:2701–2709. doi:10.1016/S0032-3861(03)00190-3
15. Olabisi O, Robeson LM, Shaw MT (1979) Polymer–polymer miscibility. Academic Press, New York
16. Utracki LA (1990) Polymer alloys and blends: thermodynamics and rheology. Hanser, Munich/New York
17. Paul DR, Newman S (eds) (1978) Polymer blends, vols 1 and 2. Academic Press, New York

18. Coleman MM, Graf JF, Painter PC (1991) Specific interactions and the miscibility of polymer blends: practical guides for predicting & designing miscible polymer mixtures. Lancaster, Technomic Pub

19. MacKnight WJ, Karasz FE, Fried JR (1978) Solid state transition behavior of blends, chap. 5. In: Paul DR, Newman S (eds) Polymer blends, vol 1. Academic Press, New York

20. Kaplan DS (1976) Structure–property relationships in copolymers to composites: molecular interpretation of the glass transition phenomenon. J Appl Polym Sci 20:2615–2629. doi:10.1002/app.1976.070201001

21. Ohno T, Nishio Y (2006) Cellulose alkyl ester/vinyl polymer blends: effects of butyryl substitution and intramolecular copolymer composition on the miscibility. Cellulose 13:245–259. doi:10.1007/s10570-005-9014-3

22. Ohno T, Nishio Y (2007) Estimation of miscibility and interaction for cellulose acetate and butyrate blends with N-vinylpyrrolidone copolymers. Macromol Chem Phys 208:622–634. doi:10.1002/macp.200600510

23. Masson J-F, Manley RSJ (1991) Cellulose/poly(4-vinylpyridine) blends. Macromolecules 24:5914–5921. doi:10.1021/ma00022a004

24. Ohno T, Yoshizawa S, Miyashita Y, Nishio Y (2005) Interaction and scale of mixing in cellulose acetate/poly(N-vinyl pyrrolidone-co-vinyl acetate) blends. Cellulose 12:281–291. doi:10.1007/s10570-004-5836-7

25. MacBriety VJ, Douglass DC (1981) Recent advances in the NMR of solid polymers. J Polym Sci Macromol Rev 16:295–366. doi:10.1002/pol.1981.230160105

26. Masson J-F, Manley RSJ (1992) Solid-state NMR of some cellulose/synthetic polymer blends. Macromolecules 25:589–592. doi:10.1021/ma00028a016

27. Miyashita Y, Kimura N, Suzuki H, Nishio Y (1998) Cellulose/poly(acryloyl morpholine) composites: synthesis by solution coagulation/bulk polymerization and analysis of phase structure. Cellulose 5:123–134. doi:10.1023/A:1009224931504

28. Hedvig P (1977) Dielectric spectroscopy of polymers. Hilger, Bristol

29. Havriliak S, Negami S (1967) A complex plane representation of dielectric and mechanical relaxation processes in some polymers. Polymer 8:161–210. doi:10.1016/0032-3861(67)90021-3

30. Cole KS, Cole RH (1941) Dispersion and absorption in dielectrics I. Alternating current characteristics. J Chem Phys 9:341–351. doi:10.1063/1.1750906

31. Nishijima Y (1970) Fluorescence methods in polymer science. J Polym Sci Part C Polym Symp 31:353–373. doi:10.1002/polc.5070310128

32. Nishio Y, Suzuki H, Sato K (1994) Molecular orientation and optical anisotropy induced by the stretching of poly(vinyl alcohol)/poly(N-vinyl pyrrolidone) blends. Polymer 35:1452–1461. doi:10.1016/0032-3861(94)90345-X

Chapter 2
Cellulosic Polymer Blends 1:
With Vinyl Polymers

Kazuki Sugimura and Yoshiyuki Nishio

Abstract This chapter reviews recent studies on blends of conventional cellulose esters (CEs), such as cellulose acetate (CA), cellulose propionate (CP), and cellulose butyrate (CB), with non-crystalline vinyl (co)polymers mainly comprising the *N*-vinyl pyrrolidone (VP) unit. The mixing behavior of the CE/VP-containing copolymer blends is seriously affected by the chain length (i.e., carbon number) and degree of substitution (DS) as to the acyl substituent of the CE component, and by the VP fraction in the copolymer component. Relatively low-substituted CEs are commonly miscible with VP-containing copolymers by virtue of hydrogen bond formation between the CE-hydroxyls and VP-carbonyls. The blend systems using CP and CB with higher DSs impart a "miscibility window" involved in the monomer composition that constitutes the copolymer component. This observation is interpretable as a result of the indirect CE–copolymer attraction driven by intra-copolymer repulsion. A convincing argument for the miscibility behavior is provided by assessing the attractive or repulsive interactions between the blend constituents in terms of the viscometric interaction parameter. In addition to the basic characterization of the blend miscibility and intermolecular interaction, we also survey some useful applications of miscible CE blends to functional films exhibiting improved adsorption, optical, or thermomechanical properties. Particular attention is paid to the molecular orientation and optical anisotropy induced in deformed films of CA/VP-methyl methacrylate polymer blends.

Keywords Blends · Cellulose ester · Interaction · Miscibility · Miscibility window · Optical films · Scale of homogeneity · Vinyl copolymer · *N*-Vinyl pyrrolidone · Viscometric interaction parameter

2.1 Introduction

Cellulose and its derivatives have been commercially available for nearly a century. Among the derivatives, organic esters of cellulose (CEs) are industrially important materials; they are utilized as packaging, coatings, excipients, molded plastics,

© The Author(s) 2017
Y. Nishio et al., *Blends and Graft Copolymers of Cellulosics*,
Biobased Polymers, DOI 10.1007/978-3-319-55321-4_2

fibers, optical films, and membranes and other separation media [1, 2]. Toward enhancing the consumption of CEs for conventional use and exploiting of new CE applications, improving or variegating their physical properties is desired to satisfy diverse requests for future use. Graft copolymerization or polymer blending of CEs can be a promising approach to this end [1, 3–5]. In any of the two techniques, the molecular-scale mixing with an appropriate second polymer component would offer an opportunity to design new polymeric materials that exhibit wide-ranging and/or synergistic functions unattainable in gross mechanical mixtures and single-component materials. In particular, miscible blending of two polymers may be practically useful to alter the original physical properties and functions of CEs at the lowest possible cost.

In reality, two different polymers rarely form a homogeneous blend, because there is little or no contribution of the combinatorial entropy of mixing to the attainment of miscibility in polymer blends [6–9]. It is still difficult to precisely predict which polymer pair will form a miscible phase. At least, however, an attractive force, such as hydrogen bonding, dipole–dipole interaction, or ionic interaction, should be required to realize the polymer–polymer miscibility, as is described in Chap. 1. In the case of cellulosic blends, residual hydroxyl groups on the glucopyranose unit can offer the specific intercomponent interactions.

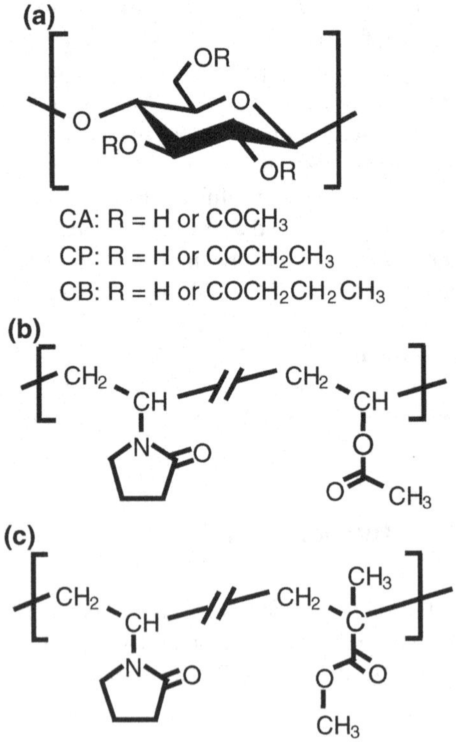
(a)
OR
O
O
RO
OR
CA: R = H or COCH3
CP: R = H or COCH2CH3
CB: R = H or COCH2CH2CH3
(b)
CH2
CH
N
O
CH2
CH
O
O
CH3
(c)
CH2
CH
N
O
CH3
CH2
C
O
O
CH3

Fig. 2.1 Structural formulae of **a** representative CEs, i.e., CA, CP, and CB, **b** P(VP-*co*-VAc), and **c** P(VP-*co*-MMA)

Concerning polymer blends of CEs, the authors' group has conducted a sequence of fundamental and practical studies for more than a decade using cellulose acetate (CA), propionate (CP), and butyrate (CB) as representative CEs (Fig. 2.1a). The counter components examined so far are categorized as two main types of polymers: biodegradable aliphatic polyesters [e.g., poly(3-hydroxybutyrate) and poly(ε-caprolactone)] and synthetic vinyl polymers. The former polymer blends are designed for biodegradable plastics, and the results will be summarized in Chap. 3. The present chapter covers work on binary blends of CEs with non-crystalline vinyl polymers. A comparative investigation of the miscibility and intermolecular interaction is summarized systematically for several systems of CE/vinyl polymer blends. Application of the miscible blends to functional optical films and membranes is also exemplified. Besides these, some related work aimed at enhancing the mechanical performance of CE blends is briefly reviewed.

2.2 Cellulose Ester Blends with *N*-Vinyl Pyrrolidone Copolymer

Cellulose acetate CA is a particularly important cellulose ester and utilized for many applications due to its desirable physical properties such as good optical clarity in film form, and a comparatively high modulus and adequate flexural and tensile strengths in fiber form. However, a trouble is that CA alone cannot be molded so easily by thermal processing, because its glass transition temperature (T_g) and melting point are fairly high, where significant thermal decomposition can take place. Therefore, in practice, a large amount of plasticizer (e.g., phthalate compound) is usually mixed with CA for the thermal molding. The use of low molecular-weight plasticizers, however, can cause fume generation in the molding process due to their volatility or decomposition. In addition, bleeding-out of plasticizers from molded CA products can be pronounced in long-term uses. Accordingly, it is of great significance to improve the original thermal property of CA by incorporation with flexible polymers as plasticizer through molecular-level interactions.

Taking into account the hydrogen-bonding formability of cellulosics, a blending partner of CEs including CA would be desired to have a proton-accepting capacity. One of the suitable polymers is poly(*N*-vinyl pyrrolidone) (PVP); in fact, it has been shown that PVP can form compatible blends with unmodified cellulose [10, 11]. Since PVP and its copolymers are widely prevailing in medical, sanitary, cosmetic, and other safety-conscious areas, the selection of vinyl copolymers containing an *N*-vinyl pyrrolidone (VP) unit (Fig. 2.1b, c) is promising for improvement in properties of the CE family. Actually, CEs are also usable in bio-related fields, e.g., as separation membranes and release-controllable excipients. The miscible incorporation with the VP-containing polymers may be expected to improve the thermal, gas permeation, and water absorption properties of CEs, and possibly provides another excellent functionality unrealized by CEs themselves; the latter effect is really found in an optical function for film use (Sect. 2.2.3).

2.2.1 Miscibility Maps as a Function of DS and Copolymer Composition

(a) Cellulose Ester/Poly(N-vinyl pyrrolidone-co-vinyl acetate) Blends

Miyashita et al. carried out the miscibility characterization of CA blends with homo- and random co-polymers comprising VP and/or vinyl acetate (VAc) units, i.e., PVP, poly(vinyl acetate) (PVAc), and poly(N-vinyl pyrrolidone-co-vinyl acetate) (P(VP-co-VAc)) (Fig. 2.1b) [12]. The CA/synthetic polymer blends were all prepared in film form by casting from mixed polymer solutions in N,N-dimethylformamide (DMF) or N,N-dimethylacetamide. On the basis of thermal transition data obtained by differential scanning calorimetry (DSC), a miscibility map (Fig. 2.2a) was completed as a function of the degree of substitution (DS) of CA and the VP fraction in P(VP-co-VAc). Figure 2.3 exemplifies a result of the DSC measurements for two blending pairs of CA/P(VP-co-VAc) corresponding to the polymer combinations marked as **A** and **B** in Fig. 2.2a; the miscibility was estimated by T_g criteria (see Sect. 1.3.1 of Chap. 1), after heat treatment of the samples above T_g of both components to equalize their thermal history. In the DSC data (Fig. 2.3a) for the blends of CA (DS = 2.95) with P(VP-co-VAc) of VP = 61 mol% (combination **A**), we can readily see a sign of poor miscibility, as is evidenced from no appreciable shift in the T_gs of both polymer components. In addition to the two distinguishable T_g signals, the melting endotherm and cold-crystallization exotherm of the CA component are both detectable for every blend at almost the same temperature positions as those for the triacetate alone. In contrast to this result, the thermograms collected in Fig. 2.3b for the pair of CA (DS = 2.70) and the same P(VP-co-VAc) (VP = 61 mol%) (combination **B**) show a miscible behavior, as indicated by a single T_g that shifts to the higher temperature side along with the increase in the CA content.

As summarized in Fig. 2.2a, the miscibility behavior of the CA/P(VP-co-VAc) series is found to be significantly affected by both the DS of CA and the copolymer composition. FT-IR and solid-state ^{13}C NMR spectroscopic measurements revealed the presence of hydrogen-bonding interactions formed between CA-hydroxyls and VP-carbonyls in the miscible blends, which was evidenced by observations of a systematic peak shift for the relevant signals of IR bands and carbon resonances [12, 13]. The role of intermolecular hydrogen-bondings in the CA–vinyl polymer miscibility was made more explicit through a pseudo complexation behavior from solutions in a poor solvent [13]. When CA and P(VP-co-VAc) solutions in tetrahydrofuran were mixed with each other, the two polymer components precipitated spontaneously to form a complex-like agglomerate, due to stronger

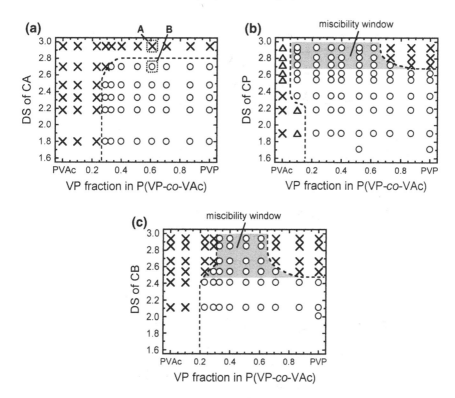

Fig. 2.2 Miscibility maps for three blend systems: **a** CA/P(VP-*co*-VAc), **b** CP/P(VP-*co*-VAc), and **c** CB/P(VP-*co*-VAc), depicted as a function of DS of CE and VP fraction in the vinyl copolymer. Symbols indicate that a given pair of CE/P(VP-*co*-VAc) is miscible (*open circle*, single T_g), immiscible (*cross*, dual T_gs), or partially miscible (*open triangle*, dual T_gs approaching each other to an appreciable degree) [Quoted with permission from [12] for (a), [14] for (b), and [15] for (c)]

attraction between the pair polymers in preference to their respective solvations. The yield of the precipitate diminished with increasing DS of CA, viz., decreasing residual hydroxyl groups on the glucopyranose residues, and with decreasing VP fraction in the vinyl copolymer. The T_g value of the complexes was always higher than that of blend films cast from solutions in DMF as good solvent. This obser-vation suggests that a high frequency of interactions combines CA with vinyl polymers intimately in their complex to seriously reduce the mobilities of the individual polymer chains.

In a similar manner to that for the CA/P(VP-*co*-VAc) system, the miscibility state in the blends of cellulose propionate (CP) [14] and butyrate (CB) [15] with P(VP-*co*-VAc) was estimated by T_g determination with DSC. The results for the two systems are summarized in Fig. 2.2b, c, respectively, by constructing the corresponding miscibility maps. Compared to the previous map of the CA series (Fig. 2.2a), the critical DS required for attainment of the miscibility of CP and CB with PVP homopolymer is appreciably lower ($\sim$2.65 for CP; $\sim$2.5 for CB). This is

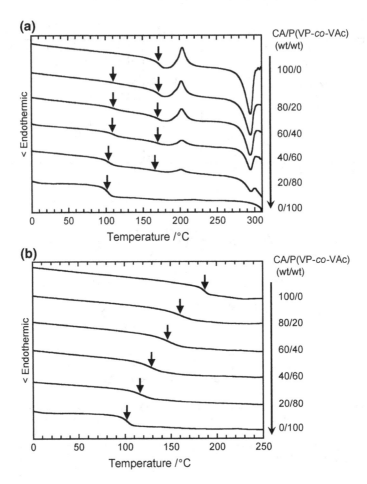

Fig. 2.3 DSC thermograms of CA/P(VP-*co*-VAc) blends: **a** data for combination **A** of CA (DS = 2.95) and P(VP-*co*-VAc) (VP:VAc = 61:39); **b** data for combination **B** of CA (DS = 2.70) and P(VP-*co*-VAc) (VP:VAc = 61:39) (see Fig. 2.2a). Arrows indicate a T_g position taken as the midpoint of a baseline shift in heat flow

because the intermolecular hydrogen-bonding interaction, as a driving force for the CE–PVP miscibility, is suppressed in frequency by steric hindrance of the comparatively bulky substituent (propionyl or butyryl side-group).

Unlike the situation in the CA series, high-substituted CPs (DS > 2.65) and CBs (DS > 2.5) make a miscible pair with some of the copolymers (not rich in VP) in spite of their immiscibility with both PVP and PVAc homopolymers (see Fig. 2.2b, c). This unique miscibility behavior, i.e., advent of a "*miscibility window*," is attributed to an indirect CE–copolymer attraction that is driven by repulsion between the two monomeric units constituting the vinyl copolymer. More concretely, since these two monomer species, VP and VAc, having mutually repellent characters were randomly combined in P(VP-*co*-VAc) by covalent bonding, the copolymers can form a miscible monophase with the CE component in the binary blends so as to reduce the strong repulsion between the comonomers. Such an explicit window never appears in the CA/P(VP-*co*-VAc) map (Fig. 2.2a). This may be interpreted as due to a stronger self-association ability of CA, which becomes pronounced when the DS exceeds ~ 2.75; the CA rather crystallizes in a cellulose triacetate II form (see Fig. 2.3a). Differing from this, no crystallizing habit is detected even for tripropionate and tributyrate samples synthesized at DS = 2.8–2.95. The lesser self-association nature of CP and CB should be advantageous to the indirect attractive interaction with the P(VP-*co*-VAc) component.

From comparison of the three maps shown in Fig. 2.2, it is astonishing afresh to find that only one difference in carbon number of the acyl substitution drastically changes the region of miscible CE/P(VP-*co*-VAc) pairings. Intriguingly, the CP system (Fig. 2.2b) provides the largest miscible region. In particular, it should be stressed that the high-substituted CPs of DS > 2.7 are miscible with P(VP-*co*-VAc) s abundant in VAc residue. At the VAc-richer compositions, the intramolecular repulsion effect in P(VP-*co*-VAc) would decline to a considerable extent. Instead, a weak interaction coming from the structural affinity between the propionyl and VAc moieties (see below) would contribute as another driving force. This is supported by T_g behavior of CP/PVAc pairs. Figure 2.4 collects T_g versus composition plots for five series of CP/PVAc blends (propionyl DS = 2.18–2.93). The blend series using CPs of DS = 2.18 and 2.35 are completely immiscible. Regarding the other blend series using CPs of DS > 2.5, however, an appreciable extent of T_g shift is

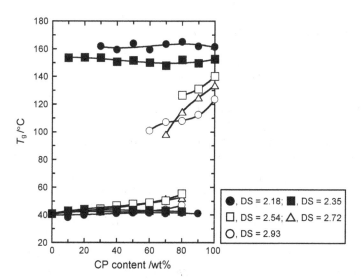

Fig. 2.4 T_g versus composition plots for five series of CP/PVAc blends. DS of CP: *filled circle*, 2.18; *filled square*, 2.35; *open square*, 2.54; *open triangle*, 2.72; *open circle*, 2.93. (Quoted with permission from [14])

detected for both of the two components at compositions of CP/PVAc = 60/40–90/10 (wt/wt). This observation indicates that a certain amount of the CP constituent is dissolved into the PVAc phase, and vice versa. Therefore, we can judge the CP (DS > 2.5)/PVAc pairs to be partially miscible. Such partial miscibility was never definable to the CA/PVAc and CB/PVAc systems irrespective of DS of the CA and CB components; the two systems always provided two invariable T_gs independent of the mixing composition. The structural affinity between the propionyl side-group (CH_3-CH_2-CO-O-C-) and the VAc unit (-(CH_2-CH(-O-CO-CH_3))-) is favorable for an interaction of dipole–dipole antiparallel alignment, which might be responsible to that partial miscibility (or better compatibility).

(b) *Cellulose Ester/Poly(N-vinyl pyrrolidone-co-methyl methacrylate) Blends*

Similar representations of the miscibility estimation based on DSC thermal analysis are given in Fig. 2.5 for two systems; a random copolymer comprising VP and methyl methacrylate (MMA) units, P(VP-*co*-MMA) (Fig. 2.1c), is combined with either CA (Fig. 2.5a) [16] or CP (Fig. 2.5b) [17]. The mapping for CB/P(VP-*co*-MMA) blends is not made in the figure, because T_gs (ca. 110–120 °C) of CBs of DS ≈ 2.5–2.9 were too close to those (ca. 100–115 °C) of P(VP-*co*-MMA)s of VP < 50 mol%. The incorporation of MMA into the vinyl polymer as a counterpart of the CE blends may be practically significant, for poly(methyl methacrylate) (PMMA) is important as optical and medical materials based on its distinguished performance and safety to living bodies.

As can be seen from the map shown in Fig. 2.5a, most of the pairs composed of CA of DS ≤ 2.6 and P(VP-*co*-MMA) of VP ≥ ∼30 mol% are miscible, whereas the other combinations of DS and VP values basically lead to an immiscible blend series. The driving factor for the miscibility attainment was ascertained to be the hydrogen-bonding formed between the CA-hydroxyls and the VP-carbonyls [16], as in the case of the CA/P(VP-*co*-VAc) system. As for CA/PMMA homopolymer blends, two independent glass transitions associated with the two components were discernible irrespective of DS of the CA used, but the T_g of the PMMA component slightly shifted to higher temperatures with increasing CA content. This suggests that PMMA shows a limited degree of miscibility with CA [16, 18]; however, the T_g of the CA component hardly shifted from the original position. Thus the CA/PMMA pairs were all judged to be substantially immiscible.

Compared to the CA system, the miscible pairing region in the CP/P(VP-*co*-MMA) map (Fig. 2.5b) expands to cover a considerably hydrophobic area of higher DS and MMA-rich composition [17]. The appearance of a miscible region of propionyl DS ≤ 2.65 and VP ≥ ∼10 mol% owes to the intercomponent hydrogen-bondings. In the upper region satisfying DS > 2.7 and VP = ca. 10–40 mol%, a clear miscibility window emerges as a result of the indirect attraction due to intramolecular repulsion in the P(VP-*co*-MMA) component. The location of the window in the side of MMA-rich compositions would owe to better affinity between propionyl and MMA moieties, which was supported by the composition-dependent shift of PMMA's T_g observed for a blend series of CP/PMMA. In comparison between the two maps shown in Figs. 2.2b and 2.5b,

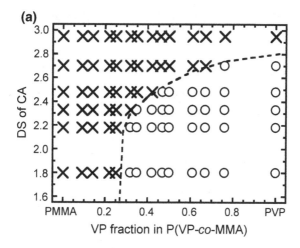

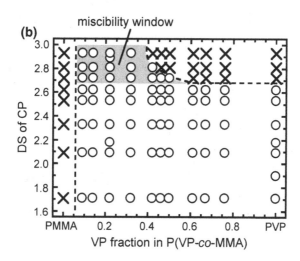

Fig. 2.5 Miscibility maps for two blend systems: **a** CA/P(VP-*co*-MMA) and **b** CP/P(VP-*co*-MMA), depicted as a function of DS of CE and VP fraction in the vinyl copolymer. The meanings of two symbols *open circle* and *cross* are the same as defined in Fig. 2.2. [Quoted with permission from [16] for (a) and [17] for (b)]

obviously, the window region for the CP/P(VP-*co*-MMA) system is narrower than that for the CP/P(VP-*co*-VAc) system. This narrowing might be ascribed to the weaker repulsion between VP and MMA units relative to that in the VP-VAc copolymer. In this connection, the contribution of such an intra-copolymer effect is made clearer from estimation of certain interaction parameters associated with the two copolymers (see Sect. 2.2.2). To make another comparison involving CA blends, the miscible region in the CA/P(VP-*co*-MMA) map (Fig. 2.5a) is also narrower than that in the CA/P(VP-*co*-VAc) map (Fig. 2.2a). This implies that the intra-copolymer effect is present latently in the CA/vinyl polymer blends, although the hydrogen-bonding interaction predominantly works to realize the respective blend miscibility.

(c) *Cellulose Mixed Ester Blends with N-Vinyl Pyrrolidone Copolymer*

From the above comparative study, it is evident that the miscibility behavior of CEs (i.e., CA, CP, and CB) with VP-containing vinyl copolymer, P(VP-*co*-VAc) or P(VP-*co*-MMA), is largely affected by a small difference in alkyl chain-length (carbon number) of the acyl substituent in the employed CE. As an expansion of the blend series using the monoester derivatives of cellulose, additional attention has been directed to the miscibility behavior of binary blends of cellulose mixed ester with the VP-containing copolymers. Traditional mixed esters, cellulose acetate propionate (CAP) and acetate butyrate (CAB), are employed here as the cellulosic component.

Figure 2.6 shows an estimation result of the blend miscibility of four different CABs (#1–4) with P(VP-*co*-VAc) [15]; the total DS of CAB used ranges from 2.4 to 2.7. CAB#1 and #2 are acetyl-rich and butyryl-rich CAB samples, respectively, and the two contain appreciable amounts of both acetyl and butyryl groups. CAB#3 and #4 are butyryl-rich CABs of total DS = 2.68 and 2.39, respectively. With regard to the CAB#3 and #4, their behavior of miscibility with the vinyl polymer (lower two data in Fig. 2.6) resembles that of CBs having comparable DSs in the total substitution (see Fig. 2.2c). The others, CAB#1 and #2, are miscible with P(VP-*co*-VAc) in a wider range of VP:VAc composition (upper two data in Fig. 2.6), when compared with the corresponding data for CA (Fig. 2.2a) and CB (Fig. 2.2c) of the same degree of acyl substitution as the total DS of the CAB considered. For instance, the blend series of CAB#2 (total DS = 2.55; acetyl:butyryl = 0.96:1.59) with P(VP-*co*-VAc) exhibits a miscibility window, as did the corresponding series of CB (DS = 2.54), but the window observed for the former is much wider than that for the latter. This expansion of the miscible range in the copolymer composition may be interpreted as due to an additional repulsion effect originating in the CAB component. That is, the cellulose mixed ester would also

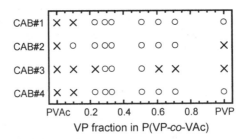

Fig. 2.6 Miscibility estimation for CAB/P(VP-*co*-VAc) blend series using four CABs: CAB#1 of total DS = 2.71 and acetyl:butyryl = 2.00:0.71; CAB#2 of total DS = 2.55 and acetyl:butyryl = 0.96:1.59; CAB#3 of total DS = 2.68 and acetyl:butyryl = 0.14:2.54; CAB#4 of total DS = 2.39 and acetyl:butyryl = 0.14:2.25. The meanings of two symbols *open circle* and *cross* are the same as defined in Fig. 2.2 (Rearranged by using data from [15], with permission)

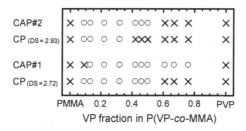

Fig. 2.7 Miscibility estimation for CAP/P(VP-*co*-MMA) blends using two CAPs (CAP#1, total DS = 2.68 and acetyl:propionyl = 0.16:2.52; CAP#2, total DS = 2.95 and acetyl:propionyl = 0.47:2.48), in comparison with the corresponding CP/P(VP-*co*-MMA) blends using CPs of DS = 2.72 and 2.93 (see Fig. 2.5b), respectively. The meanings of two symbols *open circle* and *cross* are the same as defined in Fig. 2.2 (Quoted with permission from [17])

behave as a kind of copolymer dangling two different acyl side-groups along the carbohydrate backbone. Therefore, the CAB/P(VP-*co*-VAc) system is actually taken as a copolymer/copolymer system, where the miscibility should be affected by the duplicate, intramolecular copolymer effect.

A similar deal of cellulose mixed ester as copolymer is applicable to another system of CAP/P(VP-*co*-MMA) [17]. Figure 2.7 summarizes the miscibility estimation for the two CAP/P(VP-*co*-MMA) series using CAP#1 (total DS = 2.68; acetyl:propionyl = 0.16:2.52) and CAP#2 (total DS = 2.95; acetyl:propionyl = 0.47:2.48). The corresponding data for comparable CP/P(VP-*co*-MMA) blends using CPs of DS = 2.72 and 2.93 are also shown in the figure. Owing to the additional repulsion effect originating in the CAP side, both the CAP series offer a miscibility window over a wider range of the VP:MMA copolymer composition, compared to the CP series of the corresponding DS in total.

The dual acylation leading to a mixed ester of cellulose is practically useful to improve the blend miscibility (or compatibility) with a second polymer as plasticizer. For instance, as the demonstration is described in Chap. 5, we can design a compatible system of cellulose mixed ester/flexible polymer showing an adequate melt-spinnability; thereby it will become possible to create a new type of cellulosic fiber by melt spinning.

2.2.2   Intermolecular Interaction and Homogeneity Scale

(a) *Polymer-Polymer Interaction Parameter Determined by Viscometry*

As described in Sect. 2.2.1, the CE/VP-containing copolymer combinations assume miscible or immiscible behavior according to the balance in effectiveness of the following four factors: (1) hydrogen-bonding attraction between residual hydroxyls of CE and VP-carbonyl groups of the vinyl copolymer; (2) steric hindrance of bulky side-groups to the interaction specified in (1); (3) indirect attraction via intramolecular repulsion between the comonomer units in the copolymer; and

(4) weak interaction due to structural affinity (e.g., dipole–dipole antiparallel alignment) between the ester side-group of CE (such as CH_3-CH_2-CO-O-C-) and the VAc (-(CH_2-CH(-O-CO-CH_3))-) or MMA (-(CH_2-(CH_3)C(-CO-O-CH_3))-) unit. Probably by virtue of the factors 3 and 4, the blend miscibility of CP with the VP-containing vinyl copolymer is more improved in respect of the miscible pairing number, compared to the cases using CA and CB. However, these two factors could not be directly detected by spectroscopic measurements.

To clarify the contributions of the copolymer effect and structural affinity to the miscibility attainment, attractive or repulsive interactivity between the CE and vinyl polymer constituents was quantitatively discussed in terms of polymer–polymer interaction parameters determinable by dilute solution viscometry [16, 19]. In the viscometric treatment developed by Krigbaum and Wall [20] and other groups [21, 22], the polymer–polymer miscibility can be estimated by comparison between an experimentally obtained value and an ideally calculated one of viscometric interaction parameter for dilute blend solution, i.e., b_m^{ex} and b_m^{id}, respectively. When there is a large difference between the intrinsic viscosity of both polymers ($[\eta]_1$ and $[\eta]_2$), a so-called Krigbaum-Wall interaction parameter, Δb ($= b_m^{ex} - b_m^{id}$), can be standardized, and a more effective parameter, μ, is obtained by using the following equation [22]:

$$\mu = \frac{\Delta b}{\left([\eta]_2 - [\eta]_1\right)^2} \tag{2.1}$$

If μ is positive, the polymer 1 and polymer 2 are mutually attractive and therefore the pair is taken as miscible. Contrarily, if μ is negative, the repulsive pair is considered to be immiscible. The absolute value of μ, i.e., $|\mu|$, should represent the relative strength of attractive or repulsive interaction between the two component polymer molecules. In fact, the $[\eta]$ values (2–6 dL/g) of the cellulosics and those (0.1–0.6 dL/g) of the vinyl (co)polymers are fairly far apart, and hence the standardized parameter μ is mainly used below for discussion on the interaction and miscibility between the blend constituents.

Figure 2.8 summarizes simplified miscibility maps (left) of the three blend systems of CA, CP, CB each blended with P(VP-co-VAc), with addition of the illustration in terms of μ data (right) obtained for selected polymer combinations (DS of CEs, ~ 2.7; VP:VAc of copolymer, ~ 0.5:0.5) critical to the respective systems [16, 19]. All the judgments based on μ data are actually in accordance with the miscibility map based on DSC thermal analysis.

As for a highly propionylated CP (DS = 2.72)/P(VP-co-VAc) series (Fig. 2.8b), the immiscible combination of PVP and PVAc (judged in advance by DSC) provides a larger negative μ value (-4.23×10^{-2}) than the CP/PVP (-1.02×10^{-2}) and CP/PVAc (-7.19×10^{-5}) pairs. A CB (DS = 2.67)/P(VP-co-VAc) series (Fig. 2.8c) also shows a similar relationship of repulsion (immiscibility) between the three ingredient polymer pairs. By support of the μ data, it is reasonably deduced that, due to the intense repulsion between VP and VAc segments, the P(VP-co-VAc) component can mix with the CP and CB components showing less repulsion to the comonomer units instead of the copolymer–copolymer association.

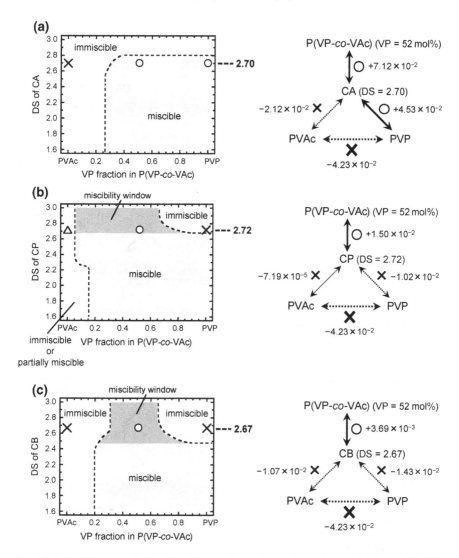

Fig. 2.8 Miscibility maps (*left*) with additional illustrations using μ data (*right*) for **a** CA/P(VP-*co*-VAc), **b** CP/P(VP-*co*-VAc), and **c** CB/P(VP-*co*-VAc) systems. The meanings of three symbols *open circle*, *cross*, and *open triangle* are the same as used in Fig. 2.2 (Quoted with permission from [19], with an adequate modification)

Another interesting observation in Fig. 2.8 is that the negative μ value for the CP (DS = 2.72)/PVAc pair is much smaller than that for the CP/PVP pair by more than two orders of magnitude. The value in $|\mu|$, 7.19×10^{-5}, for the CP/PVAc pair is also overwhelmingly small, compared with $|\mu| = 2.12 \times 10^{-2}$ for CA (DS = 2.70)/ PVAc (Fig. 2.8a) and 1.07×10^{-2} for CB (DS = 2.67)/PVAc (Fig. 2.8c). This low magnitude of μ reflects the "partially miscibility" specified in the blends of highly

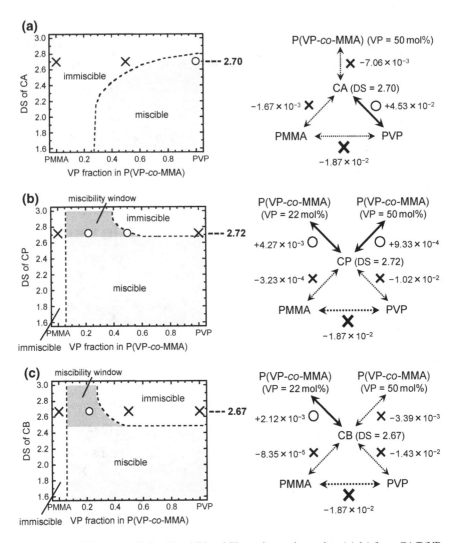

Fig. 2.9 Miscibility maps (*left*) with additional illustrations using μ data (*right*) for **a** CA/P(VP-*co*-MMA), **b** CP/P(VP-*co*-MMA), and **c** CB/P(VP-*co*-MMA) systems. The meanings of two symbols *open circle* and *cross* are the same as used in Fig. 2.2 (Rearranged by using data from [19], with permission)

substituted CPs with PVAc homopolymer (see Fig. 2.4). Such a good compatibility would lead to expansion of the miscible pairing region in the map of the CP/P(VP-*co*-VAc) system particularly to the side of VAc-richer compositions, in comparison with the cases using CA and CB.

As summarized in Fig. 2.9, μ evaluations made for CA (DS = 2.70)/P(VP-*co*-MMA) (Fig. 2.9a) and CP (DS = 2.72)/P(VP-*co*-MMA) blends (Fig. 2.9b) are also in consistency with the respective miscibility mappings based on DSC analysis. The lower order of μ (10^{-3}–10^{-4}) obtained for the CA/PMMA and CP/PMMA pairs supports the better compatibility of the CEs with PMMA. From the triangular relationship between the three ingredient polymers participating in the CP/P(VP-*co*-MMA) series, it can be ensured that the intramolecular repulsion effect of the VP-MMA copolymer gives rise to the miscibility window in the map. However, the repulsion between the VP and MMA constituents ($\mu = -1.87 \times 10^{-2}$) is weaker than the intramolecular repulsion in P(VP-*co*-VAc) (-4.23×10^{-2} shown above). This deterioration of the copolymer effect is responsible for the observation of the narrower miscible region in the CE/P(VP-*co*-MMA) maps relative to that in the CE/P(VP-co-VAc) maps (see Fig. 2.8).

Such a useful μ assessment may be applicable to complement the miscibility estimation for CB/P(VP-*co*-MMA) blends; the total mapping for this system has not been accomplished by DSC only. In Fig. 2.9c, a convincing map is constructed with the aid of μ data for the system concerned. Definitely, there appears a miscibility window in the map. The miscible range of VP:MMA composition in this window is a little bit narrow, compared with the corresponding range in the CP/P (VP-*co*-MMA) map (Fig. 2.9b).

In Fig. 2.10, the miscibility estimation for CAP#1 (total DS = 2.68; acetyl: propionyl = 0.16:2.52)/P(VP-*co*-MMA) blends is inspected in μ terms. For example, a combination of CAP#1 with P(VP-*co*-MMA) (VP = 50 mol%) provided a positive μ data of $+5.79 \times 10^{-3}$, surely supporting the situation of the polymer pair in the miscibility window of the map. Furthermore, we notice that this μ vale is considerably larger than $\mu = +9.33 \times 10^{-4}$ obtained for a comparable pair of CP (DS = 2.72)/P(VP-*co*-MMA) (VP = 50 mol%) (see Fig. 2.9b). As listed in Fig. 2.10 (right), a cellulose ester pair, CA (DS = 2.70)/CP (DS = 2.72), gave a negative μ of -8.12×10^{-3}, suggesting that a relatively strong repulsive interactivity works between the two cellulosic monoester components. From these evidences in μ terms, it is reasonably concluded that the additional intramolecular repulsion in the CAP side contributes to the expansion of the miscibility window in the mapping of the CAP#1/P(VP-*co*-MMA) blends.

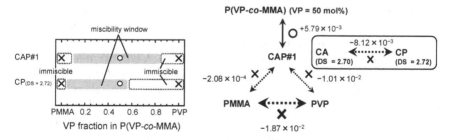

Fig. 2.10 Miscibility estimation (*left*) with additional illustrations using μ data (*right*) for binary polymer blends, CAP#1 (total DS = 2.68; acetyl:propionyl = 0.16:2.52)/P(VP-*co*-MMA) and CP (DS = 2.72)/P(VP-*co*-MMA). The meanings of two symbols *open circle* and *cross* are the same as used in Fig. 2.2 (Rearranged by using data from [19], with permission)

(b) *Scale of Homogeneous Mixing*

Physical properties of polymer blends would be greatly influenced by a level of mixing. It is therefore meaningful to quantify a scale of homogeneity in even the blends judged to be miscible. In the CE/VP-containing vinyl (co)polymer blends described thus far, the principal driving force for the miscibility attainment was the intercomponent hydrogen-bonding or the intra-copolymer repulsion. The difference in effectiveness between the driving factors should affect the size of heterogeneity (L) in the miscible blends.

As mentioned in Chap. 1, through measurements of the proton spin-lattice relaxation time in the rotating frame ($T_{1\rho}^H$) by solid-state ^{13}C NMR, we can estimate the mixing homogeneity in a scale of ^1H spin-diffusion length (= 2–5 nm) [23–25]; this dimensional scale is much smaller than that (~ 25 nm) detected by DSC thermal analysis (see Fig. 1.5). $T_{1\rho}^H$ values can be obtained by fitting the decaying carbon resonance intensities of one's concern to the following exponential equation:

$$M(\tau) = M(0)\exp(-\tau/T_{1\rho}^H) \tag{2.2}$$

where $M(\tau)$ is the magnetization intensity observed as a function of the spin-locking time τ.

Figure 2.11 displays the decay behavior of ^{13}C resonance intensities for two miscible blends of CP (DS = 1.71)/P(VP-*co*-VAc) (VP:VAc = 52:48) and CP (DS = 2.89)/P(VP-*co*-VAc) (VP:VAc = 52:48), the predominant factor contributory to the respective miscibility attainments being the intermolecular hydrogen-bonding for the former and the intramolecular repulsion in the copolymer for the latter [14]. Concerning the CP (DS = 1.71)/P(VP-*co*-VAc) blend (Fig. 2.11a), $T_{1\rho}^H$ values of the two components coincide with each other just at the midpoint (27.9 ms) between the respective original values, 20.5 ms for the CP and 35.4 ms for the copolymer. Then, using the spin-diffusion equation, Eq. (1.3), the heterogeneity size L of this blend is estimated at $L = \sim 4$ nm as the maximum. Such an equalization of $T_{1\rho}^H$s of two components was exemplified for other miscible blends of hydrogen-bonding type, not only in the CP system but also in the CA and CB systems with P(VP-*co*-VAc) or P(VP-*co*-MMA) as a mixing partner [13, 15–17]. By additional spectroscopic evidence of the interaction, it has been ensured that those blends are homogeneous at the molecular segmental level of less than a few nanometers.

In the other example using the high-substituted CP/P(VP-*co*-VAc) blend (Fig. 2.11b), there arises a serious disagreement between $T_{1\rho}^H$s of the two polymer components; that is, the respective relaxation processes proceed almost independently, indicating less cooperative spin diffusion. A similar temporal disagreement was observed for other blends situated in the miscibility window in the CP/P(VP-*co*-VAc) map (Fig. 2.2b), and this was also the case for the CP/P(VP-*co*-MMA) (Fig. 2.5b) and CB/P(VP-*co*-VAc) systems (Fig. 2.2c). By combined use of the $T_{1\rho}^H$ result and DSC data, it can be said that the scale of homogeneity in the miscible blends using highly substituted CP and CB lies between ca. 4 and 25 nm.

Fig. 2.11 Semilogarithmic plots of the decay of ^{13}C resonance intensities as a function of spin-locking time τ, for solid films of **a** CP (DS = 1.71), P(VP-*co*-VAc) (VP:VAc = 52:48), and their 50/50 blend, and **b** CP (DS = 2.89), P(VP-*co*-VAc) (VP:VAc = 52:48), and their 50/50 blend (Reproduced with permission from [14])

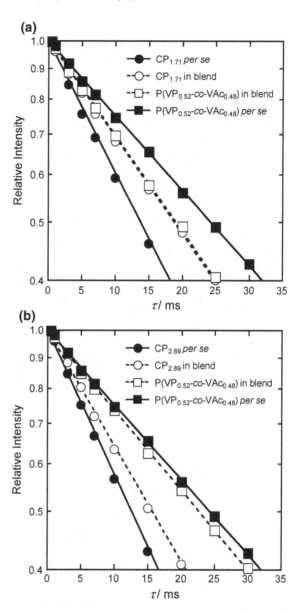

2.2.3   Application to Functional Films

In this section, we illustrate significant applications of some selected CE/vinyl (co) polymer blends to functional films. After referring to a few examples of separation membrane using CE blends, our major concern is focused on oriented films mainly of the miscible CA/P(VP-*co*-MMA) blends; this blend system has a high potential as an important optical element in the display field.

(a) *Separation Membranes*

For further functional development of CEs as membrane materials, the improvement of their adsorption property in a broad sense, such as liquid- or vapor-permeation selectivity, is required. In relation to this, Nguyen et al. examined pervaporation characteristics of CA/P(VP-*co*-VAc) blends for application as alcohol-selective membrane materials [26, 27]. According to the study, through specific interactions of VP-carbonyls in the copolymer with ethanol, the blend membranes make efficient removal of ethanol from its mixture with ethyl *tert*-butyl ether (ETBE) that is an environmentally friendly octane-value enhancer for motor fuels. In the process of ETBE production, an excess of ethanol has to be separated from the reaction system and recycled. However, the conventional separation process by distillation is quite inefficient to the ethanol/ETBE mixtures, because ethanol forms an azeotropic mixture with ETBE. The CA/P(VP-*co*-VAc) pervaporation membrane may be a potential alternative to the conventional distillation method.

Concerning gas separation membranes, He, O_2, and N_2 gas transport properties of compatible CA/PMMA blends have been measured [18]. According to this report, the blend films conditionally form a layered morphology with the surface layers rich in PMMA and the interior composed of the polymer mixture, and they show a high-selective permeability to He gas due to the unique morphology. Thus the CA/PMMA blends may be useful for a membrane material to produce a high-purity helium gas stream in a simple process.

(b) *Optical Films*

In connection with the utility of CEs as optical films to regulate or modulate polarized light in modern displays, our attention has recently been directed to the molecular orientation and optical anisotropy invited by deformation of CE-based films [5, 28]. In an elaborate work by Ohno and Nishio [28], a fluorescence polarization technique [29, 30] was utilized to obtain information about the degree and type of molecular orientation in uniaxially drawn films of the miscible CA/P (VP-*co*-MMA) blends, each film containing a slight amount of a fluorescent probe, 4,4'-bis(2-benzoxazolyl)stilbene (see Fig. 1.7a in Chap. 1). Birefringence quantification was also carried out for the drawn films, in order to estimate the state of optical anisotropy induced therein.

In Fig. 2.12a, the second moment of molecular orientation, $<\cos^2\omega>$, defined in Sect. 1.4.2 of Chap. 1, is plotted against the percentage elongation of film specimens of CA (DS = 1.80)/P(VP-*co*-MMA) (VP = 35 mol%) blends. For all the blend compositions, we can see that the value of $<\cos^2\omega>$ increases monotonically from 0.33 with increasing extent of elongation. Therefore, it follows that any of the drawn films imparts a "positive" orientation function, i.e., $f = (3<\cos^2\omega> - 1)/2 > 0$, indicating a normal trend of molecular orientation in the draw direction. However, the development of orientation becomes suppressed with increasing

Fig. 2.12 Plots of **a** $<\cos^2\omega>$ versus % elongation and **b** $<\cos^4\omega>$ versus $<\cos^2\omega>$ for drawn blends of CA (DS = 1.80)/P(VP-*co*-MMA) (VP:MMA = 35:65). (Rearranged by using data from [28] and additional data, with permission)

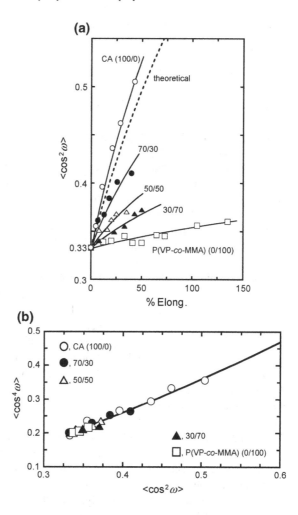

content of the vinyl polymer component; the two constituent polymers would be oriented cooperatively in the miscible blend films during the course of the uniaxial drawing process (see below).

Figure 2.12b illustrates a plot of the fourth moment $<\cos^4\omega>$ against the second moment $<\cos^2\omega>$ of molecular orientation for the CA (DS = 1.80)/P(VP-*co*-MMA) (VP = 35 mol%) blends. All the data satisfactorily fit a relationship between the two moments calculated in terms of a prolate ellipsoidal model of orientation distribution; the relationship is virtually equivalent to a theoretical one predicted by a Kratky-type affine deformation scheme [31]. Thus the type of molecular orientation in the drawn blends always follows this orthodox model, although the rate of orientation development depends on the blend composition.

Meanwhile, birefringence of an oriented polymer film, which is defined as $\Delta n = n_\parallel - n_\perp$ with a refractive index $(n_\parallel)$ parallel to the draw direction and that $(n_\perp)$ perpendicular to it, in general, varies with the degree of orientation, according to the equation:

$$\Delta n = \{(3 < \cos^2 \omega_s > - 1)\Delta n^\circ\}/2 \qquad (2.3)$$

where Δn° is an intrinsic birefringence for the perfect uniaxial orientation of the polymer chains, and $<\cos^2\omega_s>$ is the second moment of orientation for an aniso-tropic segmental unit S with a certain polarizability. Here it should be stressed that the two second moments, $<\cos^2\omega>$ and $<\cos^2\omega_s>$, obtained from the fluorescence polarization and birefringence measurements, respectively, are different in magni-tude from each other, because there is a respectable difference in size of the structural unit for orientation estimation between the two methods (see Fig. 2.13a).

Figure 2.14 compiles results of the birefringence measurements conducted for drawn films of three miscible series of CA/P(VP-co-MMA) blends prepared in different combinations of DS and VP:MMA ratio: (a) DS = 1.80 and VP: MMA = 35:65; (b) DS = 2.18 and VP:MMA = 47:53; and (c) DS = 2.48 and VP: MMA = 50:50. Drawn films of the vinyl copolymers show negative optical ani-sotropy $(\Delta n^\circ_{P(VP\text{-}co\text{-}MMA)} < 0)$, whereas the CAs (DS $\leq$ 2.75) exhibit positive optical anisotropy $(\Delta n^\circ_{CA} > 0)$ upon stretching of their films. Therefore, birefrin-gence Δn of the blends is widely controllable in the degree and polarity, by altering the DS of CA, the VP:MMA ratio in P(VP-co-MMA), and the polymer composition of blending. With a certain specific blend composition, the drawn film can behave like an optically isotropic medium even though it should be mechanically aniso-tropic due to orientation development. The critical binary composition where the blend remains a birefringence-free material shifts to the CA-rich composition side with increasing DS of the CA used and with increasing VP fraction in the P(VP-co-MMA) component (see Fig. 2.13b). As seen in Fig. 2.14b, c, some blend films rich in the vinyl copolymer can provide a negative Δn larger in magnitude than that of the respective unblended vinyl copolymers. This is because the negative contri-bution of the P(VP-co-MMA) component becomes more intensely pronounced, as a result of the cooperative orientation through hydrogen-bonding interaction with the semi-rigid CA component having a high orientation ability (see Fig. 2.13c).

A wide variation in birefringence of oriented CP blends has also been attained, using miscible combinations of CP/P(VP-co-MMA) of hydrogen-bonging type [17]. For instance, drawn films of CP (DS = 2.09)/P(VP-co-MMA) (VP = 47 mol%) can provide much larger Δn values in CP-rich compositions, compared to the corre-sponding films using a comparable CA (DS = 2.18). As an example using cellulose mixed ester, Yamaguchi et al. have studied optical birefringence of partially miscible CAP/PVAc blends that offered transparent films [32]. A zero-birefringence

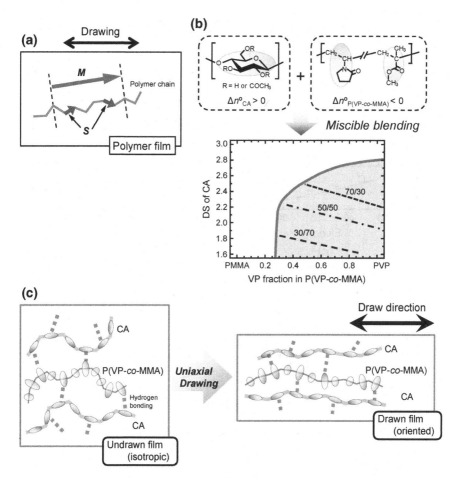

Fig. 2.13 **a** Schematic representations of two structural units *M* (the molecular axis of a fluorescent probe with a length of ∼2.5 nm) and *S* (a statistical segment of shorter length, having a specific anisotropy in polarizability). **b** Miscibility map for CA/P(VP-*co*-MMA) blends with three dotted lines, designated as 70/30, 50/50, and 30/70, which inform a critical binary composition where a given polymer blend shows a zero-birefringence nature, irrespective of the draw ratio (Rearranged by using data from [28], with permission). **c** Schematic representation of molecular orientation induced in a film of CA/P(VP-*co*-MMA) upon uniaxial drawing. In **b** and **c**, polymer chains are illustrated in terms of a sequence of the polarizability ellipsoids of the constituent monomer units

phenomenon is realized for a film prepared at a composition of CAP:PVAc = ∼50:50; i.e., the drawn specimens exhibited no birefringence irrespective of the draw ratio.

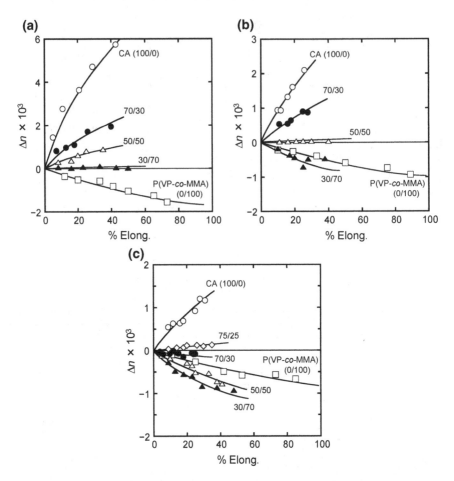

Fig. 2.14 Birefringence Δn versus % elongation for drawn films of CA/P(VP-*co*-MMA) blends prepared at different combinations of DS and VP:MMA ratio: **a** DS = 1.80 and VP: MMA = 35:65; **b** DS = 2.18 and VP:MMA = 47:53; and **c** DS = 2.48 and VP/MMA = 50:50 (Reproduced with permission from [28])

2.3   Other Prominent Systems of Cellulose Ester/Vinyl Polymer Blend

In addition to the VP-containing vinyl (co)polymers employed above, a few hydrophilic vinyl polymers are known to be miscible with a couple of conventional CEs. There are several prominent studies on thermal and/or mechanical enhancements of the miscible blends by introducing intercomponent cross-links or inorganic clays therein, as summarized below. On the other hand, miscible blends of CEs with hydrophobic vinyl (homo)polymers are still rare. To attain their intimate mixing, however, there are some attempts to construct an interpenetrating polymer network (IPN). This chemical technique is also a promising route to synergistically improve the thermal and mechanical performance and other physicochemical properties of cellulosic compositional materials.

2.3.1   Enhancement of Thermomechanical Performance

By the attainment of miscible blends of CE with amorphous vinyl polymers, the thermal (e.g., T_g), optical (e.g., orientation birefringence), and adsorption properties of CE can be improved, but the mechanical strength and heat resistance are often deteriorated. A remedy for such drawbacks in thermomechanical stability of the CE-based blends is introduction of chemical cross-linkages, as has been applied to a miscible CE/poly(vinyl phenol) (PVPh) system [33]; PVPh has potential for a secondary chemical reaction with cross-linking agent leading to the chain extension. This phenolic polymer is known to form miscible blends with a wide variety of polymers with hydrogen-bond accepting groups including commercially available CEs [34, 35]. Kelley et al. prepared polymer networks from miscible CA/PVPh and CAB/PVPh blends containing a latent formaldehyde source as cross-linker and showed improvement of the strength and thermal stability of the whole system [33].

Another method to effectively reinforce CE-based materials may be the use of layered clays as nano-filler [3, 36]. A recent successful example by the authors' group was nanoincorporation of various types of layered double hydroxides (LDHs) in a miscible blend of CA/poly(acryloyl morpholine) (PACMO) [37]. The LDHs were incorporated by solution blending of CA with a polymer/inorganic hybrid precursor prepared by polymerization of ACMO monomer containing each LDH powder. Amphiphilically modified organophilic LDHs (LDH[M]) were well exfoliated and dispersed in the cast blend films; this gave rise to an excellent effect of thermomechanical reinforcement, which was assessed by dynamic mechanical analysis (see Fig. 2.15).

2.3.2   Intimate Mixing of Cellulosic Blend by In Situ Polymerization of Vinyl Monomer

As summarized in previous reviews [3, 38, 39], highly compatible blends are obtained for some pairs of unmodified cellulose/vinyl polymer, using a suitable non-aqueous solvent and procedure for each individual case. As a noteworthy act in the '90 s, Nishio and Miyashita proposed a new route leading to unique micro-composites of cellulose [40, 41]. In situ polymerization of vinyl monomers as coagulant and/or impregnant used to form cellulose gels is an essential part of the method. Thereby, it is possible to synthesize an IPN of cellulose/vinyl polymer, where the mixing level of the two polymers can be at a few nanometers, even though the pair is originally immiscible.

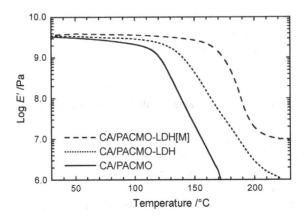

Fig. 2.15 Temperature dependence of the dynamic storage modulus E' for CA/PACMO, CA/PACMO-LDH, and CA/PACMO-LDH[M] (Quoted with permission from [37]). Their polymer compositions were unified as CA/PACMO = 50/50 (wt/wt), and the content of LDH or LDH[M] was adjusted to ~3.2 wt%. In this example, 12-hydroxysteric acid-modified LDH was use as LDH[M]

A similar technique of "in situ polymerization" is applicable to microscopic composition of CEs with vinyl polymers. Aoki et al. have successfully exemplified unique IPN composites made up of CA and PMMA [42]; the pair usually forms immiscible blends when prepared by simple solution-casting. An SH-containing CA was prepared in advance by additional esterification of CA with mercaptoacetic acid, and then the interpenetration and cross-linking between the polymer components were simultaneously attained by thiol–ene polymerization of MMA in solutions of the modified CA. The dried composites, belong to a joined type of IPN [3], exhibited a higher tensile strength and a better ductility in film form, without parallel in any film of CA, PMMA, and immiscible CA/PMMA blends.

2.4 Concluding Remarks

The present chapter surveyed recent studies on the blend miscibility of CEs mainly with vinyl copolymers comprising the VP unit. The estimation was conducted by DSC thermal analysis (T_g detection), and the results were represented in the miscibility maps as a function of DS of the CE component and VP fraction of the copolymer component. Through comparison between the maps, it was revealed that the miscibility behavior of the CEs with a given vinyl copolymer drastically changed depending on the carbon number of the acyl substituent of CE. In addition to the specific intermolecular interactions such as hydrogen bonding, the repulsive intra-copolymer effect played an important role for the miscibility attainment; the latter effect was embodied by the appearance of a "miscibility window" in the maps for the CP and CB blend systems. Attractive or repulsive interactivities between the blend-constituting polymers were also evaluated in terms of the viscometric interaction parameter. Consequently, we were able to satisfactorily comprehend the miscibility behavior of the CE/VP-containing copolymer blends.

Besides the basic characterization, we reviewed some useful studies of miscible CE blends that showed remarkably improved adsorption, optical, or thermomechanical properties toward their possible functional development. In the representative example for the miscible CA/P(VP-*co*-MMA) system, the orientation birefringence of deformed blend films was proved to be widely controllable. The result demonstrating a synergistic effect in optical property of the CE-based blends provides a suggestive idea to develop new optical media using cellulosics in the future.

References

1. Edgar KJ, Buchanan CM, Debenham JS, Rundquist PA, Seiler BD, Shelton MC, Tindall D (2001) Advances in cellulose ester performance and application. Prog Polym Sci 26: 1605–1688. doi:10.1016/S0079-6700(01)00027-2
2. Rustemeyer P (ed) (2004) Cellulose acetates: properties and applications. Wiley-VCH, Weinheim
3. Nishio Y (2006) Material functionalization of cellulose and related polysaccharides via diverse microcompositions. Adv Polym Sci 205:97–151. doi:10.1007/12_095
4. Teramoto Y (2015) Functional thermoplastic materials from derivatives of cellulose and related structural polysaccharides. Molecules 20:5487–5527. doi:10.3390/molecules 20045487
5. Sugimura K, Teramoto Y, Nishio Y (2015) Cellulose acetate. In: Kobayashi S, Müllen K (eds) Encyclopedia of polymeric nanomaterials. Springer, Berlin/Heidelberg, pp 339–347
6. Paul DR, Newman S (eds) (1978) Polymer blends, vols 1 & 2. Academic Press, New York
7. Olabisi O, Robeson LM, Shaw MT (1979) Polymer–polymer miscibility. Academic Press, New York
8. Utracki LA (1990) Polymer alloys and blends: thermodynamics and rheology. Hanser, Munich/New York
9. Coleman MM, Graf JF, Painter PC (1991) Specific interactions and the miscibility of polymer blends: practical guides for predicting & designing miscible polymer mixtures. Technomic Pub., Lancaster
10. Masson J-F, Manley RSJ (1991) Miscible blends of cellulose and poly(vinylpyrrolidone). Macromolecules 24:6670–6679. doi:10.1021/ma00025a018
11. Miyashita Y, Kimura N, Nishio Y, Suzuki H (1994) Transition behavior and phase structure of cellulose/poly(*N*-vinylpyrrolidone) composites prepared by a solution coagulation/bulk polymerization method. Kobunshi Ronbunshu 51:466–471. doi:10.1295/koron.51.466
12. Miyashita Y, Suzuki T, Nishio Y (2002) Miscibility of cellulose acetate with vinyl polymers. Cellulose 9:215–223. doi:10.1023/A:1021144827845
13. Ohno T, Yoshizawa S, Miyashita Y, Nishio Y (2005) Interaction and scale of mixing in cellulose acetate/poly(*N*-vinyl pyrrolidone-co-vinyl acetate) blends. Cellulose 12:281–291. doi:10.1007/s10570-004-5836-7
14. Sugimura K, Katano S, Teramoto Y, Nishio Y (2013) Cellulose propionate/poly(*N*-vinyl pyrrolidone-*co*-vinyl acetate) blends: dependence of the miscibility on propionyl DS and copolymer composition. Cellulose 20:239–252. doi:10.1007/s10570-012-9797-y

15. Ohno T, Nishio Y (2006) Cellulose alkyl ester/vinyl polymer blends: effects of butyryl substitution and intramolecular copolymer composition on the miscibility. Cellulose 13: 245–259. doi:10.1007/s10570-005-9014-3

16. Ohno T, Nishio Y (2007) Estimation of miscibility and interaction for cellulose acetate and butyrate blends with N-vinylpyrrolidone copolymers. Macromol Chem Phys 208:622–634. doi:10.1002/macp.200600510

17. Sugimura K, Teramoto Y, Nishio Y (2013) Blend miscibility of cellulose propionate with poly(N-vinyl pyrrolidone-co-methyl methacrylate). Carbohydr Polym 98:532–541. doi:10. 1016/j.carbpol.2013.06.045

18. Bikson B, Nelson JK, Muruganandam N (1994) Composite cellulose acetate/poly(methyl methacrylate) blend gas separation membranes. J Membr Sci 94:313–328. doi:10.1016/0376-7388(94)87041-1

19. Sugimura K, Teramoto Y, Nishio Y (2015) Insight into miscibility behaviour of cellulose ester blends with N-vinyl pyrrolidone copolymers in terms of viscometric interaction parameters. Cellulose 22:2349–2363. doi:10.1007/s10570-015-0660-9

20. Krigbaum WR, Wall FT (1950) Viscosities of binary polymeric mixtures. J Polym Sci 5: 505–514. doi:10.1002/pol.1950.120050408

21. Cragg LH, Bigelow CC (1955) The viscosity slope constant k'—ternary systems: polymer–polymer–solvent. J Polym Sci 16:177–191. doi:10.1002/pol.1955.120168208

22. Chee KK (1990) Determination of polymer–polymer miscibility by viscometry. Eur Polym J 26:423–426. doi:10.1016/0014-3057(90)90044-5

23. MacBriety VJ, Douglass DC (1981) Recent advances in the NMR of solid polymers. J Polym Sci Macromol Rev 16:295–366. doi:10.1002/pol.1981.230160105

24. Masson J-F, Manley RSJ (1992) Solid-state NMR of some cellulose/synthetic polymer blends. Macromolecules 25:589–592. doi:10.1021/ma00028a016

25. Miyashita Y, Kimura N, Suzuki H, Nishio Y (1998) Cellulose/poly(acryloyl morpholine) composites: synthesis by solution coagulation/bulk polymerization and analysis of phase structure. Cellulose 5:123–134. doi:10.1023/A:1009224931504

26. Nguyen QT, Noezar I, Clément R, Streicher C, Brueschke H (1997) Poly(vinyl pyrrolidone-co-vinyl acetate)–cellulose acetate blends as novel pervaporation membranes for ethanol–ethyl tertio-butyl ether separation. Polym Adv Technol 8:477–486. doi:10.1002/(SICI)1099-1581(199708)8:8<477:AID-PAT653>3.0.CO;2-0

27. Nguyen QT, Clément R, Noezar I, Lochon P (1998) Performances of poly (vinylpyrrolidone-co-vinyl acetate)-cellulose acetate blend membranes in the pervaporation of ethanol–ethyl tert-butyl ether mixtures: simplified model for flux prediction. Sep Purif Technol 13:237–245. doi:10.1016/S1383-5866(98)00046-X

28. Ohno T, Nishio Y (2007) Molecular orientation and optical anisotropy in drawn films of miscible blends composed of cellulose acetate and poly(N-vinylpyrrolidone-co-methyl methacrylate). Macromolecules 40:3468–3476. doi:10.1021/ma062920t

29. Nishijima Y (1970) Fluorescence methods in polymer science. J Polym Sci Part C Polym Symp 31:353–373. doi:10.1002/polc.5070310128

30. Nishio Y, Suzuki H, Sato K (1994) Molecular orientation and optical anisotropy induced by the stretching of poly(vinyl alcohol)/poly(N-vinyl pyrrolidone) blends. Polymer 35: 1452–1461. doi:10.1016/0032-3861(94)90345-X

31. Kratky O (1933) Zum deformationsmechanismus der faserstoffe, I (On the deformation mechanism of the fiber, I). Kolloid-Z 64:213–222. doi:10.1007/BF01434162

32. Yamaguchi M, Masuzawa K (2007) Birefringence control for binary blends of cellulose acetate propionate and poly(vinyl acetate). Eur Polym J 43:3277–3282. doi:10.1016/j. eurpolymj.2007.06.007

33. Gaibler DW, Rochefort WE, Wilson JB, Kelley SS (2004) Blends of cellulose ester/phenolic polymers–chemical and thermal properties of blends with polyvinyl phenol. Cellulose 11:225–237. doi:10.1023/B:CELL.0000025425.00668.de

34. Landry MR, Massa DJ, Landry CJT, Teegarden DM, Colby RH, Long TE, Henrichs PM (1994) A survey of polyvinylphenol blend miscibility. J Appl Polym Sci 54:991–1011. doi:10.1002/app.1994.070540801

35. Davis MF, Wang XM, Myers MD, Iwamiya JH, Kelley SS (1998) A study of the molecular interactions occurring in blends of cellulose esters and phenolic polymers, chap. 20. In: Heinze TJ, Glasser WG (eds) Cellulose derivatives: modification, characterization, and nanostructures. (ACS Symp Ser vol. 688) American Chemical Society, Washington DC

36. Park HM, Misra M, Drzal LT, Mohanty AK (2004) "Green" nanocomposites from cellulose acetate bioplastic and clay: effect of eco-friendly triethyl citrate plasticizer. Biomacromolecules 5:2281–2288. doi:10.1021/bm049690f

37. Yoshitake S, Suzuki T, Miyashita Y, Aoki D, Teramoto Y, Nishio Y (2013) Nanoincorporation of layered double hydroxides into a miscible blend system of cellulose acetate with poly(acryloyl morpholine). Carbohydr Polym 93:331–338. doi:10.1016/j.carbpol.2012.03.036

38. Nishio Y (1994) Hyperfine composites of cellulose with synthetic polymers, chap. 5. In: Gilbert RD (ed) Cellulosic polymers, blends and composites. Carl Hanser, Munich

39. Vigo TL (1998) Interaction of cellulose with other polymers: retrospective and prospective. Polym Adv Technol 9:539–548. doi:10.1002/(SICI)1099-1581(199809)9:9<539:AID-PAT813>3.0.CO;2-I

40. Miyashita Y, Nishio Y, Kimura N, Suzuki H, Iwata M (1996) Transition behaviour of cellulose/poly(N-vinylpyrrolidone-co-glycidyl methacrylate) composites synthesized by a solution coagulation/bulk polymerization method. Polymer 37:1949–1957. doi:10.1016/0032-3861(96)87313-7

41. Miyashita Y, Yamada Y, Kimura N, Suzuki H, Iwata M, Nishio Y (1997) Phase structure of chitin/poly(glycidyl methacrylate) composites synthesized by a solution coagulation/bulk polymerization method. Polymer 38:6181–6187. doi:10.1016/S0032-3861(97)00174-2

42. Aoki D, Teramoto Y, Nishio Y (2011) Cellulose acetate/poly(methyl methacrylate) interpenetrating networks: synthesis and estimation of thermal and mechanical properties. Cellulose 18:1441–1454. doi:10.1007/s10570-011-9580-5

Chapter 3
Cellulosic Polymer Blends 2: With Aliphatic Polyesters

Ryosuke Kusumi and Yoshikuni Teramoto

Abstract This chapter presents a review of the authors' studies on blends of cellulosic and chitinous polymers with a typical biodegradable polyester, poly (ε-caprolactone) (PCL). A miscibility map is constructed for a series of cellulose ester (CE)/PCL blends as a function of the carbon number (N) in the acyl substituent of CE and the degree of substitution (DS). The map reveals that cellulose butyrate (CB), with $N = 4$, is miscible with PCL at a comparatively lower DS, owing to a structural similarity advantage for the ester side-group of CB with a repeating unit of PCL. The melt-crystallization behavior of PCL in the miscible blends is also described, and the observed slower kinetics is interpreted in terms of a thermodynamic diluent effect of the CE component. A similar miscibility characterization is made for a comparable series of acylated chitin (Acyl-Ch)/PCL blends. The blend miscibility of the chitinous series is generally lower owing to the concurrence of N-acylation at the C2 position than that for the cellulosic series. The tensile ductility and cytocompatibility are evaluated for selected Acyl-Ch/PCL blends with different degrees of miscibility and crystallinity, by using their thermally molded and alkali-treated films. The adaptability of the chitinous blends as cell-scaffolding materials is attainable by adequately controlling the mixing state of the polymer components.

Keywords Acyl chitins · Cellulose esters · Crystallization · Cytocompatibility · Poly(ε-caprolactone) · Polymer blends · Miscibility · Tensile property

3.1 Introduction

In the past three decades, aliphatic polyesters or poly(hydroxyalkanoate)s (PHAs) produced from bacteria [1], plant-origin monomers [2], or fossil resources [3] have attracted much attention as biodegradable polymers for commodity use or as biomedical materials. However, PHAs have some disadvantages: high manufacturing costs and undesirable physical properties, including poor mechanical performances (toughness and ductility) and low thermal stability (low melting and

© The Author(s) 2017
Y. Nishio et al., *Blends and Graft Copolymers of Cellulosics*,
Biobased Polymers, DOI 10.1007/978-3-319-55321-4_3

decomposition temperatures). Different techniques have been considered to overcome the physical drawbacks of PHAs, e.g., copolymerization [4, 5], synthesis of ultrahigh molecular weight products [6], and stretching treatments [7].

A viable approach for improving the original physical properties and developing new functionalities for existing polymers is "polymer alloy" or polymer multi-composition [8]. Multicomponent polymer materials, which contain two or more different polymeric ingredients, can roughly be classified as polymer blends and copolymers with block or graft structures. The former is easily prepared at a low cost if miscible or practically compatible blend systems can be found, and the latter has the advantage of flexibility in the molecular design (see Chap. 4).

In polymer blends, miscibility (or compatibility) among the constituents is important because useful physical performances can be accomplished by structural uniformity on a microscopic level. However, immiscibility occurs so frequently that the miscibility/compatibility must be treated as an exception because an entropic contribution is significantly smaller for polymer mixing than for small molecule mixtures (solutions). The miscibility is usually observed under a precisely defined set of conditions, such as the presence of specific intermolecular interactions or an indirect driving force via intramolecular repulsion, as reviewed in Chap. 2. Miscibility attainment thus generally demands profound insights based on gathering lots of data; partly for this reason, relatively small numbers of research groups have used polymer blending to modify the performance of PHAs.

Considering the bioconformity and environmental degradability of PHAs, naturally occurring polysaccharides represented by cellulose and its relatives are of great promise as complementary materials [9]. Among cellulose derivatives, cellulose esters (CEs) have formed a practically valuable family, already finding standard acceptance for several applications, such as packing, coating, molded plastics, fibers, and optical films [10]. Even though the main chain of CEs and that of PHAs are completely different from each other, the polymer combination may be ideal because the biodegradability under adequate circumstances has also been confirmed for conventional CEs, e.g., acetate (CA), propionate (CP), and their mixed ester (CAP) [10–13]. Additionally, the PHA component can play the role of polymeric external plasticizer for the polysaccharide derivatives that show generally higher thermal-molding temperatures. Some properties of CEs can vary over a range by controlling the size of substituent and degree of substitution (DS) [14], which would affect the blend miscibility with PHAs, as demonstrated in this chapter.

In the meantime, chitin, whose molecular structure is similar to that of cellulose (but position C2 is replaced by an acetylamino group), is also abundantly available from marine crustaceans. Although the consumption of chitin for material usage is still limited unlike the situation of cellulosics, use of the amino-polysaccharide is expected to increase in biomedical materials because of the active biocompatibility [15] and antibacterial ability [16]. Blending of chitinous polymers with PHAs will invite the improvement of thermomechanical properties of each polymer and synergistic effects available for biomedical materials.

In this chapter, the authors review their systematic studies on the miscibility characterization of blends of CEs and acyl chitins (Acyl-Chs) with poly (ε-caprolactone) (PCL) as a representative PHA. The semicrystalline PCL has a high susceptibility to hydrolysis by lipases and esterases [17]. The effects of the normal acyl side-chain length and DS of CEs and Acyl-Chs are clarified by elaborating "miscibility maps" based on thermal transition data obtained by differential scanning calorimetry (DSC). We will be able to propose a working theory for miscibility attainment in relation to the partial molecular structures of the polymer ingredients. For the CE/PCL series, the melt-crystallization behavior of the PCL component is followed because some properties (including biodegradability) of the blends should be substantially affected by the development of crystallinity and an ensuing supramolecular morphology. For the Acyl-Ch/PCL series, we also focus on the cytocompatibility of selected blends for their feasibility as a cell-scaffold material.

3.2 Cellulose Ester Blends with Poly(ε-caprolactone)

Polymer blending of CEs with biodegradable PHAs was reported first by Scandola et al. [18–20] and Buchanan et al. [21–23] in the early 1990s. Both groups dealt with mainly cellulose mixed esters such as CAP and cellulose acetate butyrate (CAB), and blended them with aliphatic polyesters, e.g., poly(3-hydroxybutyrate) (PHB) [18, 19], poly(3-hydroxybutyrate-co-3-hydroxyvalerate) (PHBV) [20, 21], and linear polyesters consisting of glutaric acid and diol with various chain lengths [22, 23]. Through examination of the miscibility, Scandola's group discussed the degradability of CAP and CAB blends with the bacterial PHAs in activated sludge in connection with the phase morphology [24–26], whereas Buchanan and coworkers revealed the biodegradation behavior of blends of CAP with the dicarboxylic acid–diol polyesters in a bench-scale simulated municipal compost environment [27]. Thereafter, many related studies have been done for similar polymer blends by other researchers as well as by the two groups.

Meanwhile, Nishio et al. systematically investigated the miscibility of CEs having different normal alkyl-chain lengths with PCL [28, 29]. Although some cellulose derivatives including CAB had been suggested to be miscible with PCL [30–33] in 1970s, the relationship between the ester structure, i.e., side-chain length and DS, and the ability to form a miscible blend with PCL had been unclear. In what follows, we compile the fundamental knowledge accumulated for blends of CEs of $N = 2$–7 (N, number of carbons in the ester side-group) with PCL.

3.2.1 Dependence of Miscibility on Alkyl Side-Chain Length and DS

CEs of different side-chain lengths and DSs were prepared by a homogeneous reaction between cotton cellulose and acyl chlorides in N,N-dimethylacetamide

(DMAc)-LiCl solution. The miscibility characterization of CE/PCL blends was carried out through detection of the glass transition on the thermograms of DSC (see Chap. 1). In Fig. 3.1, DSC thermograms obtained for two blend series of cellulose valerate (CV) of DS = 3.0 and 0.72 with PCL are shown as examples of miscible and immiscible CE/PCL blends, respectively. Hereafter, acetyl, propionyl, butyryl, and valeryl DSs are coded as DS_{Ac}, DS_{Pr}, DS_{Bu}, and DS_{Va}, respectively. In Fig. 3.1a, the CV(DS_{Va} = 3.0)/PCL blends show a single glass transition temperature (T_g), which varies between the T_gs of the two constituent polymers depending on the blend composition. In addition, the crystallinity of the PCL component disappears when the CV content increases to 70 wt%. In the composition range of CV/PCL = 20/80–60/40, an exothermic peak due to cold crystallization is observed between T_g and melting temperature (T_m) of the respective samples. These are typical observations for amorphous/crystalline polymer pairs miscible in the amorphous fraction. For the CV(DS_{Va} = 0.72)/PCL blends, on the other hand, two T_gs are situated at almost the same positions as those of the homopolymers. The melting endotherm of the crystalline PCL component is detected for all the blend samples without any shift of the T_m, and no cold-crystallization exotherm of this component is discernible for any composition. Thus the CV(DS_{Va} = 0.72)/PCL pair can be taken as completely immiscible.

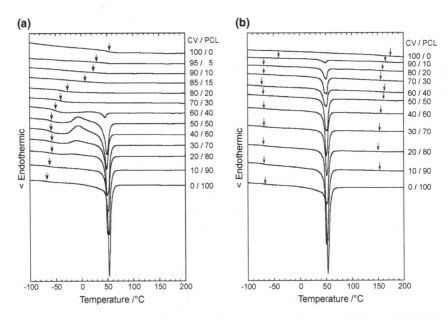

Fig. 3.1 DSC thermograms obtained for **a** CV(DS_{Va} = 3.0)/PCL and **b** CV(DS_{Va} = 0.72)/PCL blends in the second heating scans after heated to 200 °C. *Arrows* indicate a T_g position taken as the midpoint of a baseline shift appearing clearly on an enlarged scale of heat flow (reproduced with permission from [29])

Fig. 3.2 Miscibility map for different cellulose alkyl ester/PCL blends, as a function of the number of carbons in the ester side-chain and the acyl DS of the cellulosic component. *Symbols* indicate that a given pair of CE/PCL is miscible (*circle*), immiscible (*cross*), or partially miscible (*triangle*) (reproduced with permission from [29])

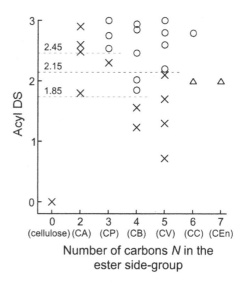

Results of the miscibility estimation for CE blends with PCL are summarized in Fig. 3.2. CP, cellulose butyrate (CB), and CV exhibit miscibility with PCL under the substitution condition of $DS_{Pr} > 2.45$, $DS_{Bu} > 1.85$, and $DS_{Va} > 2.15$, respectively. In contrast to these CEs, the most commodity-type CA is immiscible with PCL even in the highly acetylated state of $DS_{Ac} > 2.5$. Cellulose caproate (CC) and enanthate (CEn) were estimated to be miscible or partially miscible with PCL at a certain level. It is deduced that CB exhibits the highest miscibility on blending with PCL in all the cellulose alkyl esters examined.

3.2.2   Miscibility of Cellulose Mixed Ester Series

We added several samples of cellulose mixed esters for blending with PCL, to gain further insight into the effect of the side-chain structure on the miscibility with PCL. The estimations for two series of cellulose butyrate valerate (CBV) blends of $DS_{Bu}/DS_{Va} \approx 0.5/2.5$ and $DS_{Bu}/DS_{Va} \approx 2.5/0.5$ disclosed that the CBVs are both miscible with PCL. A blend series using CBV of $DS_{Bu}/DS_{Va} \approx 0.9/0.9$ was judged to be an immiscible system. These estimations are rationalized by taking into account that the DS_{Bu} of CB and DS_{Va} of CV leading to miscible blends with PCL are, respectively, >1.85 and >2.15 (Fig. 3.2).

Another CBV series of $DS_{Bu}/DS_{Va} \approx 1.5/1.5$, however, formed a single amorphous phase at every composition, as is evidenced by transition data shown in Fig. 3.3. Although both CB and CV are immiscible with PCL under the substitution condition of DS = 1.5, such a totally full-substituted CBV can be miscible, probably irrespective of the ratio of butyryl to valeryl group. In relation to this, a CA sample and a CP one, both lacking the miscibility with PCL, were transformed into

Fig. 3.3 DSC thermograms obtained for a series of CBV ($DS_{Bu}/DS_{Va} \approx 1.5/1.5$)/PCL blends. *Arrows* indicate a T_g position (reproduced with permission from [29])

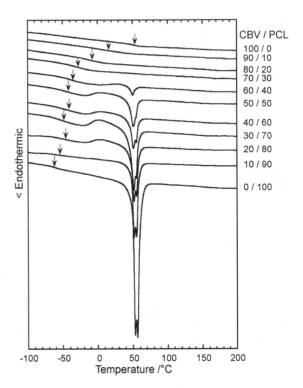

cellulose acetate valerate (CAV) ($DS_{Ac}/DS_{Va} \approx 1.0/2.0$) and cellulose propionate valerate (CPV) ($DS_{Pr}/DS_{Va} \approx 1.0/2.0$), respectively, by substituting valeryl groups for the residual hydroxyls and examined for a possible change in the miscibility behavior. The DSC analysis replied that the mixed esters were decidedly miscible with PCL, despite the DS_{Va} values of <2.15. This result implies that the critical degree of valeryl substitution required for miscibility attainment with PCL should be lowered from 2.15 by replacing the residual hydroxyl groups of CV with another acyl group. This is probably due to the decline in self-association nature of the cellulose derivative which originally possesses an OH–OH hydrogen-bonding ability. Similar inference was possible for the mixed ester derivatives having a butyryl group as one of the substituents as well. Thus a reasonable systematization was accomplished not only for the series of CE with a single substituent but also for mixed esters.

3.2.3   Miscibility Factors in CE/PCL Blends

As has been reviewed so far, CP, CB, and CV form miscible blends with PCL if the ester side-chains are introduced at DSs higher than each critical one. The mixed ester derivatives rich in butyryl or valeryl substituent show a good miscibility with

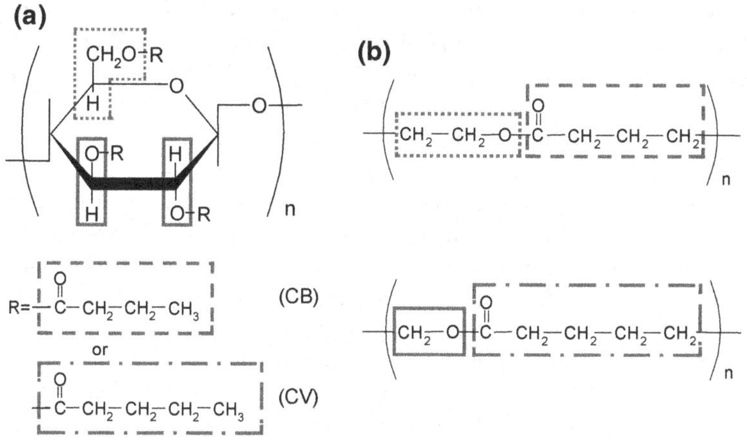

Fig. 3.4 Schematic representation of the similarity in chemical structure between the repeating unit of PCL and the butyryl and valeryl side-groups in CB and CV (reproduced with permission from [29]). Structural formulae: **a** CB and CV; **b** PCL repeating unit

PCL when the amount of the residual hydroxyls is non-dominant. For achievement of polymer/polymer miscibility, it is generally necessary to exercise some attractive interaction between the two components. For example, in blends of CE and *N*-vinyl pyrrolidone (VP)-containing vinyl polymer [34–36], the hydrogen-bonding inter-action can work between the residual hydroxyls of the CE component and the carbonyl groups of VP units of the vinyl polymer component, as a driving force for the miscibility attainment (see Chap. 2). Solid-state NMR and Fourier transform infrared (FT-IR) spectroscopy measurements for CE/PCL miscible blends, how-ever, gave no direct evidence of strong intermolecular interactions such as hydrogen bonding; thus the residual hydroxyls are never concerned in a factor of the mis-cibility between CEs and PCL. Instead, we proposed that the structural homology between the two polymer constituents would much contribute for the blend mis-cibility, considering the chemical structures of the employed CEs. As illustrated in Fig. 3.4, the butyryl ($OCOC_3H_7$) and valeryl ($OCOC_4H_9$) side-groups of CB and CV, respectively, are structurally identical with the repeating unit of PCL, if the carbons in the glucopyranose ring are taken into account. This similarity, which is lower in CP and missing in CA, may be a crucial factor, together with a possible accompanying contribution of a weak dipole–dipole interaction between the car-bonyl of the cellulose ester and that of the PCL component.

3.2.4 Crystallization Behavior

In general, mechanical strength, vapor absorption property, and biodegradability, etc. of polymer materials are much affected by development of crystallinity and supramolecular morphology therein. Therefore, it is of great significance to investigate the crystallization of the high-crystalline polyester PCL in the blends with CEs that can be miscible with the PCL component in the amorphous mixing state. In the following, we review the analysis of isothermal melt-crystallization behavior and ensuing morphology for PCL-rich blends with CB ($DS_{Bu} > 2.0$) and with CV ($DS_{Va} > 2.2$).

Figure 3.5 illustrates DSC thermograms of unblended PCL and of its blends with CB ($DS_{Bu} = 2.02$) and CB ($DS_{Bu} = 1.56$), the respective data being followed with an elapsing time (t) during the crystallization process at an isothermal crystallization temperature (T_{ic}) of 29 °C. As regards the miscible CB ($DS_{Bu} = 2.02$)/PCL blends, the exothermic curves are broadened with increasing cellulose ester content (Fig. 3.5a), indicating qualitatively that the increment in the CB content leads to decrease of the crystallization rate of the PCL component in the blends. The immiscible CB ($DS_{Bu} = 1.56$)/PCL blends for comparison do not show such a marked and systematic effect, as can be seen in Fig. 3.5b.

Using that kind of DSC isotherms, the crystallization kinetics was analyzed by the conventional Avrami equation [37]:

$$1 - X(t) = \exp(-Kt^n) \tag{3.1}$$

where $X(t)$ is the relative crystallinity at crystallization time t; K is the overall kinetic rate constant; n is the so-called Avrami exponent. The exponent n and rate constant K can be determined from the slope and intercept, respectively, of linear regression in a plot of $\log[-\ln\{1 - X(t)\}]$ versus $\log t$. The rate constant K can also be described in the following form:

$$K = \ln2/(t_{1/2})^n \tag{3.2}$$

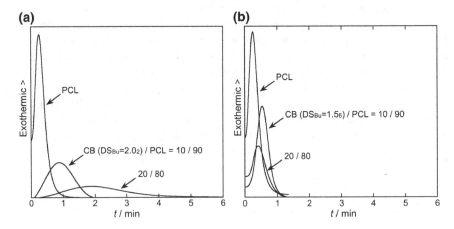

Fig. 3.5 DSC exotherms during isothermal crystallization at 29 °C after heating up to 120 °C, obtained for **a** CB($DS_{Bu} = 2.02$)/PCL and **b** CB($DS_{Bu} = 1.56$)/PCL blends with different compositions (reproduced with permission from [29])

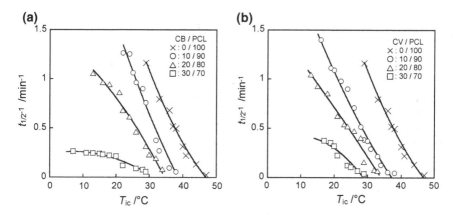

Fig. 3.6 Plots of the inverse of half-time of crystallization, $t_{1/2}^{-1}$, as a function of isothermal crystallization temperature T_{ic} for **a** CB(DS$_{Bu}$ = 2.02)/PCL and **b** CV(DS$_{Va}$ = 2.23)/PCL blends with different compositions (reproduced with permission from [29])

which includes a half-time of crystallization, $t_{1/2}$, giving $X(t) = 1/2$. Figure 3.6 shows plots of $t_{1/2}^{-1}$ versus T_{ic} for 0/100–30/70 compositions of the CB(DS$_{Bu}$ = 2.02)/PCL and CV(DS$_{Va}$ = 2.23)/PCL miscible blends. It is evident that an increase in the CB or CV content causes a remarkable diminution in the $t_{1/2}^{-1}$ value. Values of the Avrami exponent n estimated for the blends were mostly in a range of ca. 2.6–3.5, rather comparable to n data of 3.0–3.4 for plain PCL. The finding of no serious reduction in the exponent n for the blends is of significance; because this indicates that the CB and CV components never act as a heterogeneous nucleation agent (or domain) which would normally accelerate crystallization of the crystallizable polymer component from the blended melt, with a lowering of n by ~1 [38, 39]. Thus the slower crystallization observed above is explicable as due to a diluent effect of the cellulose esters that are dispersed homogeneously in the molten PCL.

Isothermally crystallized samples of the miscible CB/PCL and CV/PCL series were also examined for evaluation of the equilibrium melting temperature (T_m^{eq}) of the respective PCL crystals formed, according to the Hoffman-Weeks equation [40]:

$$T_m' = \phi T_{ic} + (1 - \phi)T_m^{eq} \tag{3.3}$$

where T_m' is an apparent melting temperature of the sample crystallized at T_{ic}; ϕ is a stability parameter of crystallization, assuming a value between 0 and 1. The estimated T_m^{eq} values were systematically lowered from 64.8 °C (neat PCL) to 59.4 °C or slightly over, with increasing content of the CB or CV component (Table 3.1). On the other hand, for all the compositions of 0/100–30/70 of any series explored, a fairly small ϕ (0.10–0.12) was obtained, suggesting that the PCL component was crystallized stably as such. Generally, the melting point depression observed for a miscible system composed of an amorphous polymer/crystalline polymer pair may be ascribed to two effects [41]: the so-called thermodynamic diluent effect [42] and a morphological effect such as lowering of the crystalline

Table 3.1 Values of equilibrium melting temperature (T_m^{eq}), nucleation factor (K_g), and surface free energy (σ_e) estimated for PCL and its blends with CB(DS_{Bu} = 2.02), CB(DS_{Bu} = 2.46), CV (DS_{Va} = 2.23), and CV(DS_{Va} = 2.85)

Samples	T_m^{eq} (°C)	$K_g \times 10^{-4}$ (K²)	$\sigma_e \times 10^5$ (J cm⁻²)
PCL	64.8	6.45	0.77
CB (DS_{Bu} = 2.02)/PCL = 10/90	60.8	9.33	1.13
20/80	60.2	9.52	1.15
30/70	59.4	10.0	1.21
CB (DS_{Bu} = 2.46)/PCL = 10/90	61.0	9.22	1.11
20/80	60.5	11.8	1.42
30/70	60.7	13.4	1.61
CV (DS_{Va} = 2.23)/PCL = 10/90	63.5	11.1	1.33
20/80	62.5	12.3	1.48
30/70	61.7	14.8	1.79
CV (DS_{Va} = 2.85)/PCL = 10/90	63.1	8.58	1.03
20/80	61.6	9.44	1.14
30/70	60.6	9.58	1.16

Reproduced with permission from [29]

order and/or size of the formed crystal. The contribution of the latter effect can be removed or minimized by analysis of the melting point depression through construction of the Hoffman-Weeks plots [43]. Therefore, the T_m^{eq}-depression found here can be taken as due to the diluent effect of the CE component that should be thermodynamically miscible with molten PCL, in agreement with the inference from the Avrami analysis.

Further insight was provided into an estimation of the surface energy of PCL lamellar crystals formed in the blends with CB and CV, basically in terms of a kinetic theory of chain-folded polymer crystallization [44, 45]. As an extension of the Turnbull-Fisher expression for nucleation [46], the crystal growth rate G may be written in the following form taking account of an amorphous/crystalline binary polymer system [39, 47–49].

$$G = v_2 G_0 \exp\left(-\Delta E_D^* / kT_{ic}\right)\exp\left(-\Delta\Phi^* / kT_{ic}\right) \tag{3.4}$$

where G_0 is a pre-exponential factor generally assumed to be constant or proportional to T_{ic}; v_2 is a volume fraction of the crystalline component; ΔE_D^* is an activation energy for the transport of crystallizing units across the molten liquid–solid interface; $\Delta\Phi^*$ is a free energy required to form a nucleus of critical size; k is the Boltzmann constant. For the present blend systems, Eq. (3.4) can be transformed to the following expression [47, 50]:

$$\alpha = \log G_0 - C\left[T_m^{eq} / (T_{ic}\Delta T)\right] \tag{3.5}$$

with

$$\alpha = (1/n)\log K - \log v_2 + 17240/[2.30R(51.6 + T_{ic} - T_g)] - (0.2T_m^{eq}\log v_2)/\Delta T \tag{3.6}$$

$$C = K_g/(2.30T_m^{eq}) = (Yb_0^2\sigma_e)/(23.0k) \tag{3.7}$$

where K_g is a nucleation factor for folded chain crystallization [44, 45]; ΔT is undercooling defined as $T_m^{eq} - T_{ic}$; b_0, which is taken as 0.414 nm in the present case of PCL [49, 51], is the distance between two adjacent fold planes; σ_e is the interfacial free energy per unit area perpendicular to the molecular chain axis; R is the gas constant. For the four miscible series shown in Table 3.1, v_2 values were determined from the density data $\rho_1 = 1.241, 1.219, 1.182,$ and 1.163 g/cm^3 for CB (DS$_{Bu}$ = 2.02), CB (DS$_{Bu}$ = 2.46), CV (DS$_{Va}$ = 2.23), and CV (DS$_{Va}$ = 2.85), respectively, and $\rho_2 = 1.094$ g/cm^3 [52] for amorphous PCL, by the following equation:

$$v_2 = (w_2/\rho_2)/(w_1/\rho_1 + w_2/\rho_2) \tag{3.8}$$

where w_1 and w_2 are weight fractions of the cellulose ester and PCL components, respectively. Y in Eq. (3.7) is a coefficient that depends on the regime of crystal growth: $Y = 4$ for regime I and $Y = 2$ for regime II [44]. Through the so-called Z test of Lauritzen [45], it was confirmed that the crystallization of PCL in all the samples proceeded according to regime II. Hence, the nucleation factor K_g involving σ_e can be evaluated by plotting α against $T_m^{eq}/(T_{ic}\,\Delta T)$ with $Y = 2$.

The result of the calculation of the folding-surface free energy σ_e is summarized in Table 3.1. Evidently, any of the miscible binary series provides a σ_e value larger than that obtained for unblended PCL, and the value tends to increase with increasing CB or CV content. The larger σ_es for the blends suggest looser fold surfaces of the PCL lamellae, despite the stable formation of the crystals in their own way. The bulky cellulose ester components would be trapped in the interfacial regions between the PCL lamellar crystals, so as to be mixed with the extruded amorphous PCL chains and possibly with the irregularly folded chain-segments of PCL as well on the lamellar surfaces.

3.2.5 Spherulite Growth Observation

The crystal growth observation for the miscible CB/PCL and CV/PCL blends with a polarized optical microscope (POM) provided additional features of the mixing state of the CE chains in the assembly of PCL lamellar crystals. The typical morphologies are shown in Fig. 3.7. During the spherulitic growth, all the blends exhibited no explicit segregation of the CE component in both intra- and extra-spherulitic regions under the microscope. The spherulites ultimately impinged

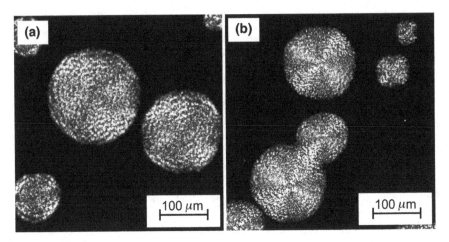

Fig. 3.7 Polarized optical micrographs of typical spherulites observed for CB/PCL and CV/PCL blends: **a** CB(DS$_{Bu}$ = 2.46)/PCL = 20/80 blend at T_{ic} = 38 °C and t = 13 min; **b** CV(DS$_{Va}$ = 2.85)/ PCL = 20/80 blend at T_{ic} = 38 °C and t = 14 min, where T_{ic} and t denote the isothermal crystallization temperature and the elapsing time, respectively (reproduced with permission from [29])

upon one another and their apparent growth stopped then. It is therefore natural to assume that the CB and CV components were incorporated in the interlamellar regions within the respective spherulites.

The spherulites exhibited a so-called Maltese-cross effect, but the texture was considerably distorted in the blends compared with that of plain PCL. In addition, the spherulites of the blends imparted a pattern composed of essentially banded extinction rings, as seen in Figs. 3.7a, b. The phenomena causing such a specific banded morphology have been interpreted by a twisted crystal model, in which the adsorption of impurity on crystal boundaries (i.e., growth faces and fold surfaces of lamellar crystals) allows the lamellae to twist around an axis of the radial growth [53]. The impurity involves a non-crystalline polymeric diluent added to the crystallizable host polymer in multicomponent polymer systems. According to the model, it may be inferred that the CE chains were trapped not only on the fold surfaces of PCL lamellae but also partly on the growth fronts in the crystallization from the molten blends.

3.3 Acylated Chitin Blends with PCL

Chitin is a representative natural polysaccharide derived from animal sources. The unique properties of chitin and its derivatives, which arise due to the presence of an amino group in the N-acetylglucosamine unit, are distinctive biological function-alities such as bioassimilability and antibacteriality. Therefore, chitin has a large potential for developing as specially functionalized materials in medical and pharmaceutical fields. However, the innate crystalline structure with multiple hydrogen bonds causes serious problems based on its poor processability and less film ductility, leading to the limited applications in tissue engineering [54–56]. To overcome the shortcomings, acylation has been studied as a modification reaction [57–64]. However, it is still not so easy to thermally mold the acylated chitin (Acyl-Ch) products alone due to the fairly high T_m (if crystallinity remains) or flowing temperature adjacent to a thermal decomposition range; actually we feel that chitinous materials seem to be susceptible to the thermal decomposition.

Here, if we apply polymer blending to chitinous materials, possibly the processability will be markedly improved without use of low molecular weight plasticizers that may be feared to badly affect the stability in long-term applications. As a counterpart constituent of chitinous blends, formerly various kinds of polymers have been proposed for achieving mutual mixing on a molecular scale [65–73]. Among them, PHAs showing good biocompatibility and environmental degradability have attracted considerable interest as candidate materials for a blending partner of chitin or chitosan [65, 74–79]. Several attempts to prepare chitinous blends with PHAs each suggested a potential route leading to biomaterials applicable to tissue engineering [78, 80] or drug delivery systems [76, 77]; these studies were based on the control of the inter-component morphology on the micrometer scale, however. Generally, it seems to be more difficult to obtain intimate mixtures of chitinous polymers and aliphatic polyesters on the nanoscopic scale.

As has been described in the above section, Nishio's group revealed that CB and CV show higher miscibility with PCL mainly due to the structural homology of the ester side-groups of the cellulosic components with a repeating unit of PCL [28, 29]. Considering the structural similarity of chitin with cellulose, the Acyl-Ch series is also expected to form miscible blends with this aliphatic polyester. PCL has in fact attracted interest as a blending partner of chitin and chitosan to obtain advanced biocompatible materials [65, 78–81]. Accordingly we first selected PCL as a counterpart constituent of Acyl-Chs to acquire blend materials exhibiting appropriate thermal processability and cytocompatibility fit to medical and sanitary uses.

In this section, we describe results of the miscibility estimation of Acyl-Ch/PCL blends [82] and compare the basic data with the corresponding one for the CE/PCL series. Subsequently, we demonstrate the availability of Acyl-Ch/PCL blend films as thermoplastic cell-scaffolding materials through evaluation of their mechanical performance and cytocompatibility [83].

3.3.1 Molecular Characterization of Acyl-Chs

Acyl-Chs with different alkyl side-chain lengths, i.e., acetate (ChA), propionate (ChP), butyrate (ChB), valerate (ChV), and caproate (ChC), were all prepared from chitin by acylation with various acid chrorides in DMAc-LiCl homogeneous solution. The chitin used was purified in advance, after isolated from crab shells.

The degree of acyl substitution (DS) of each Acyl-Ch is basically determinable by ^{1}H NMR spectroscopy in a similar manner to CEs (see Sect. 1.2.1 in Chap. 1). However, Acyl-Chs should have plural DS parameters because the reaction of acyl chlorides can occur not only with the hydroxyl groups (at C3 and C6) but also with C2–NHCOR and C2–NH$_2$ of chitin (partly deacetylated). Then, molecular characterization of the present Acyl-Chs was cautiously conducted by ^{1}H NMR, which confirmed that the acylations occurred at C3/C6 hydroxyl protons and C2 amino proton(s). It was eventually possible to determine the separate amide-DS and ester-DS values as well as their sum (total-DS).

Figure 3.8 is an example of ^{1}H NMR spectrum obtained for ChB in CDCl$_3$ solution. In the spectrum, we designate a resonance peak area derived from the methyl protons of acyl groups as **[A]**, an area of the resonance signals from the protons of glucopyranose as **[B]**, an area of the resonance of the acetamide methyl protons as **[C]**, and an area of the resonance signal from the amide N–H proton as **[D]**. The total-DS can be evaluated according to Eq. (1.1) (see Chap. 1). Values of the degree of deacetylation (DD), amide-DS, and ester-DS are determinable by the following equations:

Fig. 3.8 ^{1}H NMR spectrum of ChB of total-DS = 2.86, measured in CDCl$_3$ (reproduced with permission from [82])

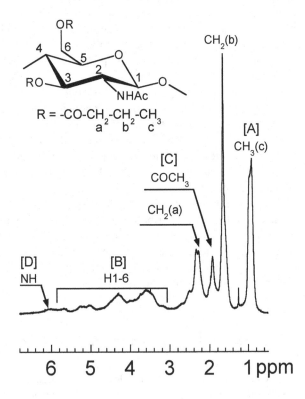

$$DD = (1 - ([\mathbf{C}]/3)/([\mathbf{B}]/7)) \times 100(\%) \tag{3.9}$$

$$\text{Amide-DS} = 2 - ([\mathbf{C}]/3 + [\mathbf{D}])/([\mathbf{B}]/7) \tag{3.10}$$

$$\text{Ester-DS} = \text{total-DS} - \text{amide-DS} \tag{3.11}$$

Hereafter, an Acyl-Ch product of total-DS = x and ester-DS = y is encoded, for example, as $\text{ChB}_{x(y)}$. Additionally remarking, amide-DS often exceeds 1.0 and therefore $x > 3$ is possible, since the replacement of N-acetyl by another acyl group can occur, besides the amidation of amine groups [82].

3.3.2   Miscibility Maps of Acyl-Ch/PCL Blends and Comparison with CE Systems

Blend films of Acyl-Chs with PCL were prepared in a wide range of compositions by casting from mixed polymer solutions in N,N-dimethylformamide. Miscibility characterization of the Acyl-Ch/PCL binary blends was made with DSC according to the guideline described in Sect. 1.3.1 of Chap. 1.

Figure 3.9 summarizes the result of miscibility estimation as a function of the number N of carbons in the acyl substituent as well as of the substitution parameters, total-DS, ester-DS, and amide-DS. ChA ($N = 2$) was immiscible with PCL even in the highly acetylated state of DS $\approx$ 3.8, as in the situation of cellulose acetate. Other Acyl-Ch products, ChP ($N = 3$), ChB ($N = 4$), ChV ($N = 5$), and ChC ($N = 6$) showed miscibility with PCL when totally high-substituted grades of them were used. A critical total-DS value for the miscibility attainment, total-DS$_{cr}$, decreased with an increase in N; however, in any of the blend series of $N = 3$–6, the total-DS$_{cr}$ is generally higher than the critical (ester-)DS observed for the corresponding CE/PCL system (see Fig. 3.2). This implies that the miscibility of

Fig. 3.9 Miscibility maps for different acylated chitin/PCL blends, as a function of the number N of carbons in the ester side-chain and the substitution parameters: **a** total-DS, **b** ester-DS, and **c** amide-DS. *Symbols* indicate that a given pair of Acyl-Ch/PCL is miscible (*circle*), immiscible (*cross*), or partially miscible (*triangle*) (reproduced with permission from [82])

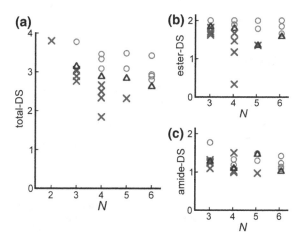

Acyl-Chs with PCL is lower compared with the case using CEs. Such a lower degree of the miscibility might be attributed partly to a steric hindrance of the bulky planar acetamide moiety to the accessibility of oxycaproyl segments of the PCL component. For the miscible CE/PCL blends, we really inferred a contribution of dipole–dipole interaction between the carbonyl of PCL and that in the normal ester side-group of CE. In the Acyl-Ch/PCL systems, similar contribution would be diminished in frequency by the steric hindrance mentioned above.

Concerning the CE/PCL systems, a structural affinity of the ester side-group of the cellulosic component with a repeating unit of PCL is a crucial factor for miscibility attainment, as stated in Sect. 3.2.3. In the Acyl-Ch series, however, the *N*-acyl substitution at C2 position cannot sufficiently contribute to such an affinity effect, while the esterification at C3/C6 positions can do, as illustrated schematically in Fig. 3.10. Then it would follow that a higher total-DS_{cr} value is required for the miscibility attainment. In support of this view, the miscibility evidently increased with an increase in ester-DS (Fig. 3.9b), but there was less correlation between the miscibility and amide-DS (Fig. 3.9c).

Here we take additional notice in comparison between the miscibility behavior of the Acyl-Ch series and that of the CE series. As shown in Fig. 3.2, any of CEs with $N = 3$–5 imparted good miscibility with PCL at DSs of ≥ 2.2, and particularly butyrate of $N = 4$ did even at a comparatively low DS of 1.9. However, cellulose caproate ($N = 6$) and enanthate ($N = 7$) having fairly longer side-chains were assumed to show a relatively lower degree of miscibility with PCL. Exceptionally, these CEs would be able to aggregate to form a specific ordered assembly, like a mesomorphic structure found for some cellulose triester [84] and chitin 3,6-*O*-diester [57] derivatives having a longer acyl substituent. With regard to the Acyl-Ch blend systems, the *N* dependence of the advent of miscibility differs from that observed for the CE systems; that is, the miscibility of the former systems increased with increasing *N* (but ≤ 6) (Fig. 3.9). Thus ChC of $N = 6$ exhibited the lowest total-DS_{cr}, signifying the highest miscibility with PCL. This indicates that the caproyl side-chains virtually identical to the repeating unit of PCL can contribute to improve the miscibility, even if it is introduced by "*N*-acylation" (Fig. 3.9c).

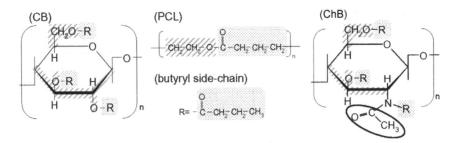

Fig. 3.10 Schematic representation of the similarity in chemical structure between the repeating unit of PCL and the butyryl side groups in CB and ChB (reproduced with permission from [82])

The self-ordering of ChC is probably prevented by the irregular molecular structure originating from the hetero-acylation.

3.3.3 Availability as Cytocompatible Flexible Films

Chitin butyrate ChB has recently been evaluated for biomedical applications utilizing its fibrous forms prepared via a dissolution-regeneration process [85–87]. As described above, the miscible blends of ChB with PCL were realized, and, therefore, they are easily expected to show thermoplasticity and serve as the scaffold of cell adhesion in various three-dimensional forms. Nevertheless, further attention should be paid to the following two points. (1) Mechanical properties such as toughness and ductility in film form of polymer blends are sometimes deteriorated for highly miscible combinations due to the abundance of amorphous phase; the drawbacks would be improved by dispersion of micro-crystallites or introduction of cross-linkages into the blends, however. (2) Surface properties of films such as hydrophilicity and accessibility by cells are core abilities for them to be a base material for cell scaffold in tissue engineering.

In what follows, we discuss the mechanical performance of films of ChB/PCL blends as representative of the Acyl-Ch/PCL series, in connection with the states of miscibility and crystallinity of the blends. Then, the cytocompatiblility of the blend films is explored through a cell adhesion test with mouse fibroblast cells, where the films are subjected to alkali-hydrolysis treatment.

(a) *Fundamental Information on Test Samples*

In sampling of films for the cell-adhesion and mechanical tests, we selected three ChB/PCL blends at a fixed composition of 50:50 in weight; they were in different miscibility states, i.e., of immiscible (IM), partially miscible (PM), and miscible (M) blends. Major structural parameters of the employed ChBs are listed in Table 3.2. The blends obtained once in film form by solution casting were each molded into a more flat film (0.1 mm thick) by hot pressing at 180 °C. After that, the molded films were conditioned at 23 °C and 50% RH for 2 weeks.

Table 3.2 Characterization of the tested ChB samples

Sample code	Total-DS	Ester-DS	Amide-DS	DD (%)	T_g^a (°C)	Miscibility with PCL[b]
$ChB_{2.34(1.19)}$	2.34	1.19	1.15	39	134	IM
$ChB_{2.93(1.76)}$	2.93	1.76	1.16	46	125	PM
$ChB_{3.45(1.86)}$	3.45	1.86	1.59	72	97	M

Reproduced with permission from [83]

[a]Determined from DSC thermograms recorded in the 2nd heating process at a scanning rate of 20 °C/min

[b]Abbreviations M, PM, and IM indicate miscible, partially miscible, and immiscible, respectively

Fig. 3.11 Visual appearance of the hot-press molded (180 °C) films of ChB$_{3.45(1.86)}$ and IM, PM, and M blends (ChB/PCL = 50/50 in weight) (reproduced with permission from [83]). The blends were prepared using three ChBs listed in Table 3.2

Figure 3.11 shows visual appearance of the film specimens of M, PM, and IM blends, together with that of a specimen of ChB$_{3.45(1.86)}$ as reference. All these films were visually homogeneous and exhibited moderate transparency. The highest light transmittance was observed for the unblended ChB$_{3.45(1.86)}$ film (78% at 550 nm). With regard to the three blend films of 50/50 composition, the transmittance (at 550 nm) decreased in the order of miscibility, i.e., estimated as 52 (for M), 39 (for PM), and 24% (for IM). The habitual crystal development of PCL is, more or less, responsible to the deterioration in transparency of these blend films. A yellow-brownish color observed for the IM sample might be derived from a lower thermoplasticity of the ChB component (ChB$_{2.34(1.19)}$) that forms a phase-separated domain in the blend.

The PCL crystallinity in the three ChB/PCL films was estimated in DSC (1st heating scan), and the result is listed in Table 3.3. All the blend samples as well as a neat PCL as control exhibited a melting endotherm and no cold-crystallization exotherm, indicating sufficient development of PCL crystals in the storage period for conditioning. The crystallinity index X_c of each sample was calculated per net weight of PCL constituent, by the following equation:

$$X_c = \Delta H_m / \left(0.5 \Delta H_m^\circ\right) \tag{3.12}$$

Table 3.3 Thermal and crystallographic property of ChB/PCL blends (50/50 in weight) and PCL, measured for their film specimens stored at 23 °C for 2 weeks after thermal molding

Sample code	T_m (°C)	ΔH_m (J/g)	$X_{c(PCL)}$[a] (%)	D (nm)
M blend	58	20.6	30.3	25.4
PM blend	59	23.4	34.5	11.6
IM blend	61	32.2	47.4	33.7
PCL	65	92.5	68.0	29.4

Reproduced with permission from [83]
[a]% crystallinity per net weight of PCL constituent

where ΔH_m and $\Delta H°_m$ (= 136 J/g) [52] are the heats of fusion of the sample and of 100-% crystalline PCL, respectively, and 0.5 is the weight fraction of the PCL component. As shown in the table, X_c and T_m of the blends decreased with increasing degree of miscibility.

Table 3.3 also contains the average crystallite size D of the PCL component, which were estimated from the half-height width of a wide-angle X-ray diffraction (WAXD) peak at $2\theta = 21.4°$ derived from the (110) plane using the Scherrer's formula:

$$D = 0.94\lambda/(\beta \cos \theta) \tag{3.13}$$

where λ, β, and θ are the wavelength of the X-ray used (=0.154 nm), the half-height width of the peak, and the Bragg angle, respectively. The size D of PCL crystals decreased by blending the polyester with any of the ChBs, but it was specifically smaller in the PM blend rather than in the M blend showing a lower X_c. For the molded PM sample, it can be assumed that relatively finer crystallites (microcrystallites) of PCL were uniformly dispersed in the amorphous ChB/PCL matrix.

To further elucidate the phase structure in several tens of nanometers for the M and PM samples, a mixing scale of the blend components was evaluated using solid-state NMR. The ^1H spin-lattice relaxation times (T_1^H) of neat ChB and PCL were ~ 1.39 and ~ 0.95 s, respectively. As for the M and PM blends, the values of the two components were very close to each other, whereas such an attunement was rightly not observed for the IM blend. As mentioned in Sect. 1.3.2 (Chap. 1), this result indicates that the two constituent polymers in the M and PM blends reside together in a range where the mutual ^1H-spin diffusion is permitted over a period of the respective homogenized T_1^H (1.00–1.13 s). The maximum diffusive path length L, which is given by Eq. (1.3) in Sect. 1.3.2, was estimated as ca. 43–52 nm for both blends. Homogeneity in the amorphous regions of the M blend has already been ensured being on the scale of a couple of tens nanometers by T_g detection with DSC (Sect. 3.3.2). For the PM blend, it can be deduced that the homogeneity is in an intermediate range between approximately 20 and 50 nm.

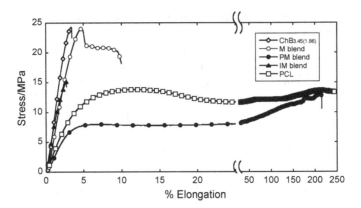

Fig. 3.12 Stress-strain curves for ChB$_{3.45(1.86)}$, IM, PM, and M blends (ChB/PCL = 50/50 in weight), and PCL (reproduced with permission from [83]). The tensile behavior was examined at 23 °C for dumbbell-type test pieces (JIS K6251: length and width of the narrow portion, 12 and 2 mm, respectively) with a thickness of 0.1 mm (see Fig. 3.11). The strain rate and span length were 0.75 mm/min and 15 mm, respectively

(b) *Tensile Mechanical Properties*

Figure 3.12 shows stress–strain curves for the film specimens of ChB/PCL = 50/50, ChB$_{3.45(1.86)}$, and neat PCL mentioned above. Numerical data determined by the tensile test (conducted at 23 °C) are shown in Table 3.4. The original mechanical properties of the constituting polymers are in contrast with each other: stiff and brittle in ChB of relatively high strength and modulus; flexible and ductile in PCL of lower modulus. The results for the blend samples reveal that there is essentially no improvement in the IM blend due to poor interfacial adhesion between the blend components. For the M and PM blends, we find a remediation in the brittleness, i.e., an increase in the ductility of ChB by blending with PCL. However, the tensile properties of these blend films were not simple arithmetic average of the two original materials, and were remarkably influenced by the degree of miscibility and crystalline morphology. For instance, the elongation at rapture for the PM blend film was estimated to be >200%, which was extraordinarily larger than that for the M blend. The PM blend also exhibited a much lower Young's modulus, compared with the other blends.

The contrastive results for the M and PM blends can be explained as follows. As regards the M blend, there is substantially no interface between domains on a scale of >20 nm, at least in the amorphous phase ($T_g \approx -30$ °C [83]). Therefore, cooperative extensional deformation of the chitinous and PCL molecules is possible with a weak dipolar interaction between the two constituents. However, the deformation would be restricted to some degree by the presence of relatively larger PCL crystals in the miscible amorphous phase (see Table 3.3), which results in a moderate improvement of the film ductility. In the PM blend exhibiting the highest ductility, there is a good balance between the crystallite dimension ($D \approx 12$ nm)

Table 3.4 Tensile mechanical properties of ChB$_{3.45(1.86)}$, ChB/PCL blends (50/50 in weight) in various miscibility states, and PCL, measured for their film specimens stored at 23 °C for 2 weeks after thermal molding

Sample code	Young's modulus (MPa)	Yield stress (MPa)	Elongation at rupture (%)	Stress at rupture (MPa)
ChB$_{3.45(1.86)}$	631 (20)	–	3.5 (0.3)	24.3 (1.6)
M blend	445 (16)	16.9 (0.2)	28.1 (3.0)	15.0 (1.8)
PM blend	170 (3)	7.9 (0.1)	215 (26)	12.7 (1.5)
IM blend	425 (8)	–	2.7 (0.3)	15.2 (0.4)
PCL	285 (16)	13.9 (0.7)	>333	>13.9

Reproduced with permission from [83]. Numbers shown in parentheses are standard deviations

and the size of amorphous domains (20–50 nm). This mixing state in the amorphous regions would allow the ChB and PCL constituents to modestly interact, but not interfere with the flexibility of PCL molecules. The miniaturized PCL crystallites would act as an effective physical cross-linker and/or nanofiller in the deformed blend film.

Thus a striking difference in the mechanical property between the ChB/PCL blends arises according to the degrees of miscibility and crystal development; interestingly, this is solely based on the alteration of DS of the ChB component.

(c) *Cytocompatibility after Alkaline Treatment*

First we examined an alkaline hydrolysis treatment for the film samples of M, PM, and IM blends, to improve the surface hydrophilicity and accessibility by cells. Figure 3.13 displays field-emission scanning electron microscope (FE-SEM) images of the blend films, which were observed before and after the treatment with 2 M NaOH at 37 °C for 48 h. Before the treatment, the films showed an essentially smooth surface with trace lines originating from the surface shape of Teflon substrates for thermal molding. Regarding water wettability, the as-prepared M, PM, and IM films provided high values of contact angle, $88.0 \pm 6°$, $85.0 \pm 2.7°$, and $88.5 \pm 1.5°$, respectively, for a drop of distilled water; there appeared no noticeable difference between the three values. Attenuated total reflection (ATR)-FTIR measurements revealed that, in the IM and M films, the PCL concentration in the surface region (0.5–1.2 μm in depth) was somewhat higher.

After the alkaline treatment, the IM blend film became too brittle to remain the initial shape, but the other two blends maintained their original film forms to the naked eye. The weight loss values were 0.83, 14.2, and 76.0% for the M, PM, and IM samples, respectively. As shown in Fig. 3.13, the surface morphology of the alkali-treated films was basically rough and exhibited different geometries depending on the miscibility and crystallinity. On the surface of the treated M film (Fig. 3.13a), we can see some radial textures, which might be the spherulitic textures exposed by an etching effect accompanied by slight elimination of both the

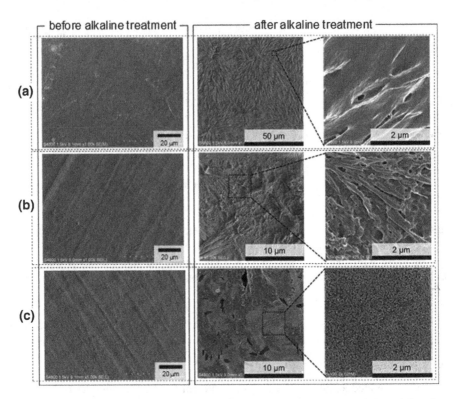

Fig. 3.13 FE-SEM images of **a** M, **b** PM, and **c** IM blends (the original ChB/PCL weight ratio, 50/50), observed before and after alkaline treatment (reproduced with permission from [83]). The alkaline hydrolysis was carried out at 37 °C in a test tube containing 2-M NaOH aqueous solution under rotary shaking (100 rpm). After 48 h, the specimens were washed thoroughly with ethanol, and then dried in vacuo at 23 °C until their weights reached constancy

amorphous PCL and chitinous components. This view was supported by the microscopic ATR-FTIR spectrum that indicated no change in the surface polymer composition before and after the alkaline hydrolysis.

The PM film exposed to the alkali solution exhibits relatively random asperity over a wide area of the surface (Fig. 3.13b). As compared by the enlarged SEM images of the M and PM blends, a more marked eluviation can be seen in the latter one. In ATR-FTIR measurement for the alkali-treated PM blend, strong bands assigned to chitin-OH and amide I, and a weaker band of PCL-C=O were observed. This result indicates that regeneration of "chitin" manifested itself in the surface region of the treated PM sample through the almost complete removal of the butyryl moiety of ChB and an appreciable elution of the PCL component. The IM blend shows a further random and inhomogeneous surface morphology with aggregation of granular particles (Fig. 3.13c), as evidenced by the enlarged SEM data. From additional data by ATR-FTIR, which was similar to that noted for the PM blend, the surface region of the treated IM sample was found to be mainly composed of chitin.

Meanwhile, the contact angle of a water drop to the hydrolyzed M and PM films was appreciably smaller (74.7 ± 0.9° and 67.2 ± 1.4°, respectively) relative to the values observed before the alkaline treatment; this indicates an improvement in surface hydrophillicity by the treatment.

Next, to evaluate cell adhesion and proliferation, mouse L929 fibroblast cells were seeded on the blend films. In Fig. 3.14, FE-SEM micrographs demonstrate

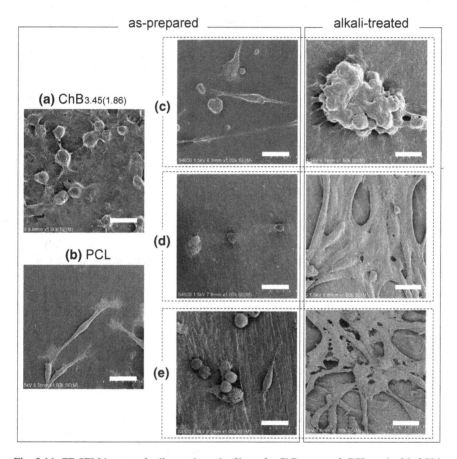

Fig. 3.14 FE-SEM images of cell growth on the films of **a** ChB$_{3.45(1.86)}$, **b** PCL, and **c** M, **d** PM, and **e** IM blends (the original ChB/PCL weight ratio for the blends, 50/50) (reproduced with permission from [83]). The images of the blend films **c–e** include the ones taken after the alkaline treatment. Scale bars denote 20 μm. L929 fibroblast cells derived from mouse were first seeded onto the sterilized film specimens at a population density of 1.0×10^5 cells/mL, and then cultured on the specimens in Eagle's minimal essential medium containing supplemental mixture at 37 °C in a humidified atmosphere of 95% air and 5% CO_2. After cultivation for 48 h, the films were rinsed twice with phosphate buffered saline to remove cells weakly adhered on the surface. Subsequently, the specimens were soaked in the 2.5 wt% glutaraldehyde solution at 4 °C for 12 h, and then dehydrated with 10-mm exposures to successive 30, 50, 70, 90, and 100% concentrations of ethanol

morphologies of the cells that were cultured for 48 h on the as-prepared films of $ChB_{3.45(1.86)}$, PCL, and the three blends and, as well, on the alkali-treated films of the blends. For the alkali-treated blend films except the M blend, it is observed that the cell adherences to the surfaces are densely distributed, in contrast to the situation without alkaline treatment where many cells are still round on the films. This observation provides an evidence of the anchoring ligands extending to help support the cells on the surfaces of the alkali-treated PM and IM films.

To sum up, the crucial factors that promotes the cell growth on these blend films are (i) the surface exposure of the chitinous component, which not only improves hydrophilicity of the films but also offers active sites for the growth, by alkaline-treatment and (ii) the high porosity that provides more structural space for cell accommodation to make an efficient exchange of nutrient and metabolic waste between the scaffold and environment. Taking into consideration the balance of the specific ductile nature and the cytocompatibility after alkaline treatment, the PM blend is the most appropriate candidate as thermoplastic scaffolding materials capable of wide application.

3.4 Concluding Remarks

In this chapter, the authors reviewed basic and application studies on the blends of acylated cellulose (i.e. CE) and chitin (i.e. Acyl-Ch) with biodegradable polyester PCL. First a miscibility map of the CE/PCL series was presented as a function of the number N of carbons in the normal acyl substituent as well as of DS; the map revealed that CB ($N = 4$) can form miscible blends with PCL at a comparatively lower DS. The miscibility attainment was attributed reasonably to the partial structural similarity of the ester side-group of the CE component with the repeating unit of PCL. Crystalline development of the PCL component in the miscible blends was also analyzed quantitatively; this analysis is important from the viewpoint of application of the blends as solid materials. The systematic understanding of the miscibility and crystallinity will also be helpful for designing CE-*gtaft*-PCL copolymers so as to diversify their molecular aggregation structure (see Chap. 4).

The second study extended for the Acyl-Ch/PCL series suggested that the miscibility in the chitinous blends was generally lowered by the *N*-acylation at the C2 position, compared with the case of the cellulosic series in perspective of the total DS. For the chitinous blend series, the availability as a thermoplastic cell-scaffolding material was demonstrated by the comparative examination using film samples of chitin butyrate (ChB)/PCL different in the degrees of miscibility and crystallinity. The result serves as a good example showing the significance of the minute characterization and control of the mixing state of ingredients in blend materials.

Polysaccharides from natural resources are much attractive as a key component for developing highly functionalized materials with biological and environmental conformity. Studies along the line of the present work will be continued and the results are, hopefully, useful in new designing of the multicomponent systems based on polysaccharides at various structural levels covering the sizes from molecule to fibril.

Acknowledgements The review in this chapter is based on the authors' studies that have been carried out in Professor Y. Nishio's laboratory of Kyoto University. The authors would like to express their sincerest gratitude for his excellent navigation and invaluable suggestions. The authors are also grateful to many colleagues in the laboratory for their kind help in various ways.

References

1. Doi Y (2002) Biotechnology: unnatural biopolymers. Nat Mater 1:207–208. doi:10.1038/nmat783
2. Fukushima K, Kimura Y (2006) Stereocomplexed polylactides (Neo-PLA) as high-performance bio-based polymers: their formation, properties, and application. Polym Int 55:626–642. doi:10.1002/pi.2010
3. Rainbolt EA, Washington KE, Biewer MC, Stefan MC (2015) Recent developments in micellar drug carriers featuring substituted poly(ε-caprolactone)s. Polym Chem 6:2369–2381. doi:10.1039/C4PY01628A
4. Huang MH, Li S, Vert M (2004) Synthesis and degradation of PLA-PCL-PLA triblock copolymer prepared by successive polymerization of ε-caprolactone and DL-lactide. Polymer 45:8675–8681. doi:10.1016/j.polymer.2004.10.054
5. Matsusaki H, Abe H, Doi Y (2000) Biosynthesis and properties of poly (3-hydroxybutyrate-co-3-hydroxyalkanoates) by recombinant strains of *Pseudomonas* sp. 61-3. Biomacromolecules 1:17–22. doi:10.1021/bm9900040
6. Kusaka S, Iwata T, Doi Y (1998) Microbial synthesis and physical properties of ultra-high-molecular-weight poly (R)-3-hydroxybutyrate. J Macromol Sci Pure Appl Chem A35:319–335. doi:10.1080/10601329808001980
7. Tanaka T, Fujita M, Takeuchi A, Suzuki Y, Uesugi K, Ito K, Fujisawa T, Doi Y, Iwata T (2006) Formation of highly ordered structure in poly[(R)-3-hydroxybutyrate-co-(R)-3-hydroxyvalerate] high-strength fibers. Macromolecules 39:2940–2946. doi:10.1021/ma0527505
8. Utracki LA (1990) Polymer alloys and blends: thermodynamics and rheology. Hanser, Munich/New York
9. Nishio Y (2006) Material functionalization of cellulose and related polysaccharides via diverse microcompositions. Adv Polym Sci 205:97–151. doi:10.1007/12_095
10. Edgar KJ, Buchanan CM, Debenham JS, Rundquist PA, Seiler BD, Shelton MC, Tinball D (2001) Advances in cellulose ester performance and application. Prog Polym Sci 26:1605–1688. doi:10.1016/S0079-6700(01)00027-2
11. Buchanan CM, Gardner RM, Komarek RJ (1993) Aerobic biodegradation of cellulose acetate. J Appl Polym Sci 47:1709–1719. doi:10.1002/app.1993.070471001
12. Gardner RM, Buchanan CM, Komarek R, Dorschel D, Boggs C, White AW (1994) Compostability of cellulose acetate films. J Appl Polym Sci 52:1477–1488. doi:10.1002/app.1994.070521012
13. Glasser WG, McCartney BK, Samaranayake G (1994) Cellulose derivatives with low degree of substitution. 3. The biodegradability of cellulose esters using a simple enzyme assay. Biotechnol Prog 10:214–219. doi:10.1021/bp00026a011
14. Teramoto Y (2015) Functional thermoplastic materials from derivatives of cellulose and related structural polysaccharides. Molecules 20:5487–5527. doi:10.3390/molecules20045487

15. Jayakumar R, Menon D, Manzoor K, Nair SV, Tamura H (2010) Biomedical applications of chitin and chitosan based nanomaterials—A short review. Carbohydr Polym 82:227–232. doi:10.1016/j.carbpol.2010.04.074

16. Dutta PK, Tripathi S, Mehrotra GK, Dutta J (2009) Perspectives for chitosan based antimicrobial films in food applications. Food Chem 114:1173–1182. doi:10.1016/j.foodchem.2008.11.047

17. Tokiwa Y, Suzuki T (1977) Hydrolysis of polyesters by lipases. Nature 270:76–78. doi:10.1038/270076a0

18. Scandola M, Ceccorulli G, Pizzoli M (1992) Miscibility of bacterial poly(3-hydroxybutyrate) with cellulose esters. Macromolecules 25:6441–6446. doi:10.1021/ma00050a009

19. Ceccorulli G, Pizzoli M, Scandola M (1993) Effect of a low molecular weight plasticizer on the thermal and viscoelastic properties of miscible blends of bacterial poly (3-hydroxybutyrate) with cellulose acetate butyrate. Macromolecules 26:6722–6726. doi:10.1021/ma00077a005

20. Lotti N, Scanola M (1992) Miscibility of bacterial poly(3-hydroxybutyrate-co-3-hydroxyvalerate) with ester substituted celluloses. Polym Bull 29:407–413. doi:10.1007/BF00944838

21. Buchanan CM, Gedon SC, White AW, Wood MD (1992) Cellulose acetate butyrate and poly (hydroxybutyrate-co-valerate) copolymer blends. Macromolecules 25:7373–7381. doi:10.1021/ma00052a046

22. Buchanan CM, Gedon SC, Pearcy BG, White AW, Wood MD (1993) Cellulose ester-aliphatic polyester blends: the influence of diol length on blend miscibility. Macromolecules 26:5704–5710. doi:10.1021/ma00073a027

23. Buchanan CM, Gedon SC, White AW, Wood MD (1993) Cellulose acetate propionate and poly(tetramethylene glutarate) blends. Macromolecules 26:2963–2967. doi:10.1021/ma00063a048

24. Tomasi G, Scandola M (1995) Blends of bacterial poly(3-hydroxybutyrate) with cellulose acetate butyrate in activated sludge. J Macromol Sci Pure Appl Chem A32:671–681. doi:10.1080/10601329508010280

25. Scandola M (1995) Polymer blends based on bacterial poly(3-hydroxybutyrate). Can J Microbiol 41:310–315. doi:10.1139/m95-202

26. Gilmore DF, Fuller RC, Schneider B, Lenz RW, Lotti N, Scandola M (1994) Biodegradability of blends of poly(3-hydroxybutyrate-co-3-hydroxyvalerate) with cellulose acetate esters in activated sludge. J Environ Polym Degrad 2:49–57. doi:10.1007/BF02073486

27. Buchanan CM, Boggs CN, Dorschel DD, Gardner RM, Komarek RJ, Watterson TL, White AW (1995) Composting of miscible cellulose acetate propionate-aliphatic polyester blends. J Environ Polym Degrad 3:1–11. doi:10.1007/BF02067788

28. Nishio Y, Matsuda K, Miyashita Y, Kimura N, Suzuki H (1997) Blends of poly(ε-caprolactone) with cellulose alkyl esters: effect of the alkyl side-chain length and degree of substitution on miscibility. Cellulose 4:131–145. doi:10.1023/A:1018475520600

29. Kusumi R, Inoue Y, Shirakawa M, Miyashita Y, Nishio Y (2008) Cellulose alkyl ester/poly(ε-caprolactone) blends: characterization of miscibility and crystallization behaviour. Cellulose 15:1–16. doi:10.1007/s10570-007-9144-x

30. Brode GL, Koleske JV (1972) Lactone polymerization and polymer properties. J Macromol Sci Chem A6:1109–1144. doi:10.1080/10601327208056888

31. Olhoft GV, Eldred NR, Koleske JV (1972) Compositions of nitrocellulose and cyclic ester polymers. US Patent, 3642507

32. Hubbell DS, Cooper SL (1977) The physical properties and morphology of poly-ε-caprolactone polymer blends. J Appl Polym Sci 21:3035–3061. doi:10.1002/app.1977.070211117

33. Koleske JV (1978) Blends containing poly(ε-caprolactone) and related polymers. In: Paul DR, Newman S (eds) Polymer blends, vol 2. Academic Press, New York, pp 369–389

34. Ohno T, Yoshizawa S, Miyashita Y, Nishio Y (2005) Interaction and scale of mixing in cellulose acetate/poly(N-vinyl pyrrolidone-co-vinyl acetate) blends. Cellulose 12:281–291. doi:10.1007/s10570-004-5836-7
35. Ohno T, Nishio Y (2007) Estimation of miscibility and interaction for cellulose acetate and butyrate blends with N-vinylpyrrolidone copolymers. Macromol Chem Phys 208:622–634. doi:10.1002/macp.200600510
36. Sugimura K, Katano S, Teramoto Y, Nishio Y (2013) Cellulose propionate/poly(N-vinyl pyrrolidone-co-vinyl acetate) blends: dependence of the miscibility on propionyl DS and copolymer composition. Cellulose 20:239–252. doi:10.1007/s10570-012-9797-y
37. Avrami M (1939) Kinetics of phase change. I General theory. J Chem Phys 7:1103–1112. doi:10.1063/1.1750380
38. Takahashi T, Nishio Y, Mizuno H (1987) Crystallization behavior of polybutene-1 in the anisotropic system blended with polypropylene. J Appl Polym Sci 34:2757–2768. doi:10.1002/app.1987.070340811
39. Nishio Y, Hirose N, Takahashi T (1990) Crystallization behavior of poly(ethylene oxide) in its blends with cellulose. Sen'i Gakkaishi 46:441–446. doi:10.2115/fiber.46.10_441
40. Hoffman JD, Weeks JJ (1962) Rate of spherulitic crystallization with chain folds in polychlorotrifluoroethylene. J Chem Phys 37:1723–1741. doi:10.1063/1.1733363
41. Nishi T, Wang TT (1975) Melting point depression and kinetic effects of cooling on crystallization in poly(vinylidene fluoride)-poly (methylmethacrylate) mixtures. Macromolecules 8:909–915. doi:10.1021/ma60048a040
42. Scott RL (1949) The thermodynamics of high polymer solutions. V. Phase equilibria in the ternary system: polymer 1-polymer 2-solvent. J Chem Phys 17:279–284. doi:10.1063/1.1747239
43. Morra BS, Stein RS (1982) Melting studies of poly(vinylidene fluoride) and its blends with poly(methyl methacrylate). J Polym Sci Polym Phys Ed 20:2243–2259. doi:10.1002/pol.1982.180201207
44. Hoffman JD, Frolen LJ, Ross GS, Lauritzen JI (1975) On the growth rate of spherulites and axialites from the melt in polyethylene fractions: regime I and regime II crystallization. J Res Natl Bur Stand Sect A Phys Chem 79A:671–699. doi:10.6028/jres.079A.026
45. Lauritzen JI, Hoffman JD (1973) Extension of theory of growth of chain-folded polymer crystals to large undercoolings. J Appl Phys 44:4340–4352. doi:10.1063/1.1661962
46. Turnbull D, Fisher JC (1949) Rate of nucleation in condensed systems. J Chem Phys 17:71–73. doi:10.1063/1.1747055
47. Boon J, Azcue JM (1968) Crystallization kinetics of polymer-diluent mixtures. Influence of benzophenone on the spherulitic growth rate of isotactic polystyrene. J Polym Sci A-2 Polym Phys 6:885–894. doi:10.1002/pol.1968.160060508
48. Martuscelli E, Silvestre C, Gismondi C (1985) Morphology, crystallization and thermal behaviour of poly(ethylene oxide)/poly(vinyl acetate) blends. Makromol Chem 186:2161–2176. doi:10.1002/macp.1985.021861020
49. Ong CJ, Price FP (1978) Blends of poly(ε-caprolactone) with poly(vinyl chloride). I. Morphology. J Polym Sci C Polym Symp 63:45–58. doi:10.1002/polc.5070630108
50. Martuscelli E (1984) Influence of composition, crystallization conditions and melt phase structure on solid morphology, kinetics of crystallization and thermal behavior of binary polymer/polymer blends. Polym Eng Sci 24:563–586. doi:10.1002/pen.760240809
51. Chatani Y, Okita Y, Tadokoro H, Yamashita Y (1970) Structural studies of polyesters. III. Crystal structure of poly-ε-caprolactone. Polym J 1:555–562. doi:10.1295/polymj.1.555
52. Khambatta FB, Warner F, Russel T, Stein RS (1976) Small-angle X-ray and light scattering studies of the morphology of blends of poly(ε-caprolactone) with poly(vinyl chloride). J Polym Sci Polym Phys Ed 14:1391–1424. doi:10.1002/pol.1976.180140805
53. Keith HD, Padden FJ, Russell TP (1989) Morphological changes in polyesters and polyamides induced by blending with small concentrations of polymer diluents. Macromolecules 22:666–675. doi:10.1021/ma00192a027

54. Jayakumar R, Prabaharan M, Nair SV, Tamura H (2010) Novel chitin and chitosan nanofibers in biomedical applications. Biotechnol Adv 28:142–150. doi:10.1016/j.biotechadv.2009.11.001

55. Khor E, Lim LY (2003) Implantable applications of chitin and chitosan. Biomaterials 24:2339–2349. doi:10.1016/S0142-9612(03)00026-7

56. Muzzarelli RAA (2009) Chitins and chitosans for the repair of wounded skin, nerve, cartilage and bone. Carbohydr Polym 76:167–182. doi:10.1016/j.carbpol.2008.11.002

57. Teramoto Y, Miyata T, Nishio Y (2006) Dual mesomorphic assemblage of chitin normal acylates and rapid enthalpy relaxation of their side chains. Biomacromolecules 7:190–198. doi:10.1021/bm050580y

58. Yang BY, Ding Q, Montgomery R (2009) Preparation and physical properties of chitin fatty acids esters. Carbohydr Res 344:336–342. doi:10.1016/j.carres.2008.11.006

59. Hirano S, Ohe Y (1975) A facile N-acylation of chitosan with carboxylic anhydrides in acidic solutions. Carbohydr Res 41:C1–C2. doi:10.1016/s0008-6215(00)87046-9

60. Kurita K, Sannan T, Iwakura Y (1977) Sudies on chitin, 3. Preparation of pure chitin, poly(N-acetyl-D-glucosamine), from water-soluble chitin. Macromol Chem Phys 178:2595–2602. doi:10.1002/macp.1977.021780910

61. Nishi N, Noguchi J, Tokura S, Shiota H (1979) Studies on chitin. I. Acetylation of chitin. Polym J 11:27–32. doi:10.1295/polymj.11.27

62. Fujii S, Kumagai H, Noda M (1980) Preparation of poly(acyl)chitosans. Carbohydr Res 83:389–393. doi:10.1016/s0008-6215(00)84553-x

63. Kaifu K, Nishi N, Komai T (1981) Preparation of hexanoyl, decanoyl, and dodecanoylchitin. J Polym Sci Polym Chem 19:2361–2363. doi:10.1002/pol.1981.170190921

64. Kurita K, Chikaoka S, Kamiya M, Koyama Y (1988) Studies on chitin. 14. N-Acetylation behavior of chitosan with acetyl chloride and acetic anhydride in a highly swelled state. Bull Chem Soc Jpn 61:927–930. doi:10.1246/bcsj.61.927

65. Honma T, Senda T, Inoue Y (2003) Thermal properties and crystallization behaviour of blends of poly(ε-caprolactone) with chitin and chitosan. Polym Int 52:1839–1846. doi:10.1002/pi.1380

66. Ko MJ, Jo WH, Kim HC, Lee SC (1997) Miscibility of chitosans/polyamide 6 blends. Polym J 29:997–1001. doi:10.1295/polymj.29.997

67. Kubota N, Konaka G, Eguchi Y (1998) Characterization of blend films of chitin regenerated from porous chitosan with poly(vinyl alcohol). Sen'i Gakkaishi 54:212–218. doi:10.2115/fiber.54.4_212

68. Lee YM, Kim SH, Kim SJ (1996) Preparation and characteristics of beta-chitin and poly(vinyl alcohol) blend. Polymer 37:5897–5905. doi:10.1016/s0032-3861(96)00449-1

69. Miyashita Y, Yamada Y, Kimura N, Nishio Y, Suzuki H (1995) Preparation and miscibility characterization of chitin/poly(N-vinyl pyrrolidone) blends. Sen'i Gakkaishi 51:396–399. doi:10.2115/fiber.51.8_396

70. Miyashita Y, Sato M, Kimura N, Nishio Y, Suzuki H (1996) An effect of deacetylation of chitin on the miscibility of chitin poly(vinyl alcohol) blends. Kobunshi Ronbunshu 53:149–154. doi:10.1295/koron.53.149

71. Miyashita Y, Yamada Y, Kimura N, Suzuki H, Iwata M, Nishio Y (1997) Phase structure of chitin poly(glycidyl methacrylate) composites synthesized by a solution coagulation bulk polymerization method. Polymer 38:6181–6187. doi:10.1016/s0032-3861(97)00174-2

72. Miyashita Y, Kobayashi R, Kimura N, Suzuki H, Nishio Y (1997) Transition behavior and phase structure of chitin/poly(2-hydroxyethyl methacrylate) composites synthesized by a solution coagulation bulk polymerization method. Carbohydr Polym 34:221–228. doi:10.1016/s0144-8617(97)00110-0

73. Nishio Y, Koide T, Miyashita Y, Kimura N, Suzuki H (1999) Water-soluble polymer blends with partially deacetylated chitin: a miscibility characterization. J Polym Sci B Polym Phys 37:1533–1538. doi:10.1002/(sici)1099-0488(19990701)37:13<1533:aid-polb19>3.0.co;2-l

74. Ikejima T, Inoue Y (2000) Crystallization behavior and environmental biodegradability of the blend films of poly(3-hydroxybutyric acid) with chitin and chitosan. Carbohydr Polym 41:351–356. doi:10.1016/s0144-8617(99)00105-8

75. Ikejima T, Yagi K, Inoue Y (1999) Thermal properties and crystallization behavior of poly (3-hydroxybutyric acid) in blends with chitin and chitosan. Macromol Chem Phys 200:413–421. doi:10.1002/(sici)1521-3935(19990201)200:2<413:aid-macp413>3.0.co;2-q

76. Mi FL, Lin YM, Wu YB, Shyu SS, Tsai YH (2002) Chitin/PLGA blend microspheres as a biodegradable drug-delivery system: phase-separation, degradation and release behavior. Biomaterials 23:3257–3267. doi:10.1016/s0142-9612(02)00084-4

77. Mi FL, Shyu SS, Lin YM, Wu YB, Peng CK, Tsai YH (2003) Chitin/PLGA blend microspheres as a biodegradable drug delivery system: a new delivery system for protein. Biomaterials 24:5023–5036. doi:10.1016/s0142-9612(03)00413-7

78. Sarasam A, Madihally SV (2005) Characterization of chitosan-polycaprolactone blends for tissue engineering applications. Biomaterials 26:5500–5508. doi:10.1016/j.biomaterials.2005.01.071

79. Senda T, He Y, Inoue Y (2002) Biodegradable blends of poly(ε-caprolactone) with α-chitin and chitosan: specific interactions, thermal properties and crystallization behavior. Polym Int 51:33–39. doi:10.1002/pi.793

80. Wan Y, Wu H, Cao XY, Dalai S (2008) Compressive mechanical properties and biodegradability of porous poly(caprolactone)/chitosan scaffolds. Polym Degrad Stab 93:1736–1741. doi:10.1016/j.polymdegradstab.2008.08.001

81. Chen B, Sun K, Ren T (2005) Mechanical and viscoelastic properties of chitin fiber reinforced poly(ε-caprolactone). Eur Polym J 41:453–457. doi:10.1016/j.eurpolymj.2004.10.015

82. Sugimoto M, Kawahara M, Teramoto Y, Nishio Y (2010) Synthesis of acyl chitin derivatives and miscibility characterization of their blends with poly(ε-caprolactone). Carbohydr Polym 79:948–954. doi:10.1016/j.carbpol.2009.10.014

83. Hashiwaki H, Teramoto Y, Nishio Y (2014) Fabrication of thermoplastic ductile films of chitin butyrate/poly(ε-caprolactone) blends and their cytocompatibility. Carbohydr Polym 114:330–338. doi:10.1016/j.carbpol.2014.08.028

84. Takada A, Fujii K, Watanabe J, Fukuda T, Miyamoto T (1994) Chain-length dependence of the mesomorphic properties of fully decanoated cellulose and cellooligosaccharides. Macromolecules 27:1651–1653. doi:10.1021/ma00084a057

85. Draczynski Z, Bogun M, Mikolajczyk T, Szparaga G, Krol P (2013) The influence of forming conditions on the properties of the fibers made of chitin butyryl-acetic copolyester for medical applications. J Appl Polym Sci 127:3569–3577. doi:10.1002/app.37784

86. Muzzarelli RAA, Guerrieri M, Goteri G, Muzzarelli C, Armeni T, Ghiselli R, Cornelissen M (2005) The biocompatibility of dibutyryl chitin in the context of wound dressings. Biomaterials 26:5844–5854. doi:10.1016/j.biomaterials.2005.03.006

87. Pant HR, Kim HJ, Bhatt LR, Joshi MK, Kim EK, Kim JI, Abdal-hay A, Hui KS, Kim CS (2013) Chitin butyrate coated electrospun nylon-6 fibers for biomedical applications. Appl Surf Sci 285(Part B):538–544. doi:10.1016/j.apsusc.2013.08.089

Chapter 4
Cellulosic Graft Copolymers

Yoshikuni Teramoto and Ryosuke Kusumi

Abstract Graft copolymerization of cellulosics has been widely investigated to modify the polymer molecules and alter the surface properties of the bulk materials. While practical advantages for the graft copolymerization were found in the surface modifications, basic and systematic studies on the cellulosic molecular grafting were not conducted very well throughout the past century. One reason for the lesser amount of systematization work was the difficulty of elaborate synthesis involved in the controlled polymerization and product isolation and characterization for obtaining the target copolymers. However, since 2000, serious efforts have been made to tackle the formulation of the structure–property relationships at molecular and supramolecular levels and the material functionalization based on those elucidations. In this chapter, the authors review the progress in copolymerization and product isolation, molecular characterization, general thermal transition scheme, thermal treatment effect on the supramolecular structure development, molecular dynamics, and orientation characteristics, mainly for cellulose ester-*graft*-aliphatic polyesters. Examples of material functionalization include controlled biodegradation and modulation of optical birefringence for molded films of the graft copolymers.

Keywords Biodegradability · Cellulose esters · Films · Functionalization · Graft copolymers · Optical property · Poly(hydroxy alkanoate)s · Structural characterization · Thermal property

4.1 Introduction

In the field of material science and technology of cellulose and other polysaccharides, graft copolymerization is generally known as a practical way to modify the molecules and alter the bulk-surface property [1, 2]. In contrast to extensive structural and physical studies for basic cellulose derivatives such as simple ethers and esters [3], systematic works dealing with morphologies, thermal transition behaviors, mechanical properties, and the like for cellulosic molecular graft

© The Author(s) 2017   75
Y. Nishio et al., *Blends and Graft Copolymers of Cellulosics*,
Biobased Polymers, DOI 10.1007/978-3-319-55321-4_4

copolymers have been restricted during the past century. One reason for the limitation is the synthetic aspect that includes the difficulties of controlled polymerization and product isolation. Nonetheless, since cellulose possesses plenty of hydroxyl groups [three per anhydroglucopyranose unit (AGU)], the molecule should essentially have a high reactivity. Thus, it is absolutely suitable for graft copolymerization if adequate polymerization systems are selected. In order to accomplish a fruitful material functionalization of cellulosic graft copolymers, efforts have been made to elucidate the detailed relationship between the copolymer composition and properties over the past dozen years or so.

Alongside the case for unmodified cellulose, the relatively good solubility of cellulose derivatives in common organic solvents makes it possible to perform grafting reactions in appropriate homogeneous systems. Cellulose acetate (CA) is one of the most important cellulose ester (CE) derivatives. Since it was reported in the mid 1990s that CAs with a degree of substitution (DS) of <2.5 are conventionally biodegradable [4, 5], new material functionalization of CA by grafting gathered momentum in many chemical industries and in agroindustrial, sanitary, and bio-related fields. As for the monomer ingredient for grafting onto CA, the use of aliphatic hydroxy acids or cyclic esters is a promising approach from a viewpoint of conformity with the environment. Graft copolymerization of CA is also an alternative way to overcome some problems by using low-molecular-weight "external" plasticizers; it achieves an "internal plasticization" of the semi-rigid cellulosic trunk polymer.

In this chapter, the authors review significant examples that clarify the relationship between structure and material properties of cellulosic graft copolymers. First, the authors provide information about how to synthesize and characterize the molecular structures, mainly with a focus on CE-*graft*-aliphatic polyesters, for which there have been appreciable data accumulations since 2000. Then, the dependence of thermal transition and relaxation behaviors on the molecular composition is rationalized. We also demonstrate some examples addressing the control of supramolecular structures and the regulation of molecular and segmental orientations for fabrication of functional materials. Attractive material functionalities include controlled biodegradability and optical birefringence characteristics.

4.2 Graft Copolymers with Aliphatic Polyester Side-Chains

Aliphatic polyesters, or poly(hydroxyl alkanoate)s (PHAs), are well known as biodegradable polymers. The research field has been vigorous since the end of the last century. Their higher-molecular-weight grades have been synthesized by polycondensation of hydroxy acids [6] and ring-opening polymerization of cyclic esters such as lactides and lactones [7]. Even though the possibility of the polycondensation was already recognized in 1930s, corresponding to the early time of polymer synthetic chemistry, its practical applications saw the light in 1990s; for it was necessary to exercise ingenuities such as the direct polycondensation in solution with a continuous dehydration [6] and the solid-phase polymerization [8]. Meanwhile, the ring-opening polymerization proceeds via the "coordination-insertion" mechanism [7] in a similar manner to living polymerization at the initial stage of the reaction, by using a tin octoate catalyst and hydroxyl initiators such as higher alcohol. Any of the polymerization mechanisms for PHAs is applicable for the graft copolymerization with cellulose derivatives, because the residual hydroxyl groups of cellulosic trunk polymers can be the initiating points for the polymerization.

4.2.1   Synthesis and Molecular Characterization

Teramoto and Nishio prepared a series of CA-*graft*-poly(lactic acid) (PLA) by three methods of grafting [9]; i.e., (1) copolycondensation of lactic acid, (2) ring-opening copolymerization of L-lactide in solution, and (3) ring-opening copolymerization similar to the method (2), but in bulk. The latter two give the grafted side-chains made up of poly(L-lactide) (PLLA). Applying these three different methods provides the graft copolymers over a wide range of compositions. The polycondensation (1) and ring-opening polymerization in solution (2) proceed relatively slowly and both produce the copolymers with the finely controlled lengths of short grafted chains, although the monomer conversion efficiencies are relatively low. Figure 4.1 exemplifies the control of compositions by the reaction mechanisms and conditions, where the structural parameters were explained in Chap. 1 and will briefly be mentioned in the ending of this subsection. On the other hand, the rapid bulk

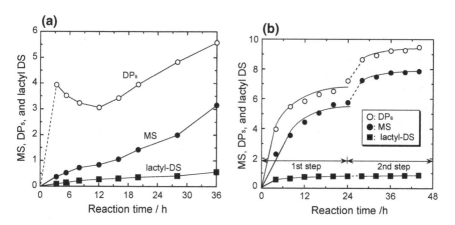

Fig. 4.1 Examples of composition control for CA$_{2.15}$-*graft*-PLA by reaction methods and time: **a** method 1 (polycondensation) and **b** method 2 (ring-opening polymerization in solution) (Reproduced with permission from [9])

ring-opening polymerization (3) uses a good solubility of CA in L-lactide and essentially yields the copolymers stoichiometrically incorporated with the in-fed monomer. This procedure has also been adopted for the synthesis of other series of CE-*graft*-PHA using cellulose butyrate (CB) as a trunk and other lactones such as (R,S)-β-butyrolactone (BL), δ-valerolactone (VL), and ε-caprolactone (CL) as monomers for graft chains [10]. These lactones are generally good solvents for the CEs as well. To achieve a further diversity of molecular architectures of the respective graft series, ester-DS of CEs as a starting polymer can also be varied, which was intended to control the intramolecular density of grafts introduced. Accordingly, we can investigate the effect of the mutual miscibility of the cellulosic trunk and grafts on physical properties, e.g., by comparing a miscible pair of CB/PCL with immiscible CA/PCL (Chap. 3).

In order to obtain the graft copolymers to warrant a quantitative discussion for structure–property relationships, isolation of pure graft copolymers is indispensable. Pure copolymers can basically be separated from the crude products as regenerated solids by a reprecipitation technique. Using gel permeation chromatography (GPC), we can check whether or not we accomplish the removals of reaction solvent (if any), unreacted monomer, and homo-polymer (or oligomer) produced by the polycondensation or by initiation of the ring-opening polymerization from hydroxyl initiators (mainly trace amount of water in the graft polymerization systems). For completion of removing them by reprecipitation, we should consider to use not only single-component non-solvent (methanol, etc.) for the grafted products, but also mixed ones such as methanol/toluene [9]. Several cycles of dissolving in a good solvent and reprecipitating and washing with a non-solvent allow extracting the impurities with number average molecular weight of $\leq 10,000$; if the molecular weight of polymeric by-products is higher than that, it is almost impossible to purify the graft copolymer products anymore.

The important parameters characterizing the graft copolymers (CE-*graft*-PHAs) are the molar substitution (MS) and the oxyalkanoyl-DS, defined as an average number of introduced oxyalkanoyl units and that of hydroxyl groups substituted by oxyalkanoyl units, respectively, per AGU of CE, which can usually be determined by ^{1}H NMR spectroscopy, as illustrated in Fig. 4.2 for a CA-*graft*-PLA. The oxyalkanoyl DS is equivalent to the DS_{graft} defined in Sect. 1.2.1 of Chap. 1. In ^{1}H NMR spectra, we designate the resonance peak area derived from the methyl protons of ester groups of CE as **A**, an area of the resonance signals from the protons of oxyalkanoyls as **B**, and the area from the terminal protons of oxyalkanoyls as **C**; then the MS, oxyalkanoyl-DS, and degree of polymerization of side-chain (DP_s) can be calculated by the following Eqs. (4.1), (4.2) and (4.3).

$$MS = (\text{ester-DS})\left(\frac{(\mathbf{B}+\mathbf{C})/n}{\mathbf{A}/3}\right) \tag{4.1}$$

$$\text{oxyalkanoyl-DS} = (\text{ester-DS})\left(\frac{\mathbf{C}/n}{\mathbf{A}/3}\right) \tag{4.2}$$

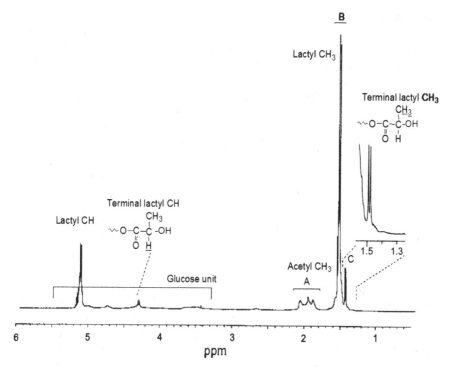

Fig. 4.2 ^{1}H NMR spectrum of a CA-*graft*-PLA for which acetyl-DS = 2.15, MS = 7.86, DP_s = 8.93, and lactyl-DS = 0.88 (Reproduced with permission from [9])

$$DP_s = \frac{MS}{oxyalkanoyl\text{-}DS} \qquad (4.3)$$

where n in Eqs. (4.1) and (4.2) is equal to 2 or 3, denoting the number of protons of the species selected for the calculations: $n = 3$ for lactic acid and BL (CH$_3$); $n = 2$ for VL (C$_\alpha$H$_2$) and CL (C$_\gamma$H$_2$). Although the signals from the terminal protons can be detected for PLA (see Fig. 4.2) and also for poly(3-hydroxybutyrate) (PHB), the polymerized product of BL, the terminal signals are not separated from the internal ones of PVL and PCL. For the latter two grafts, the apparent DP_s should be calculated by DP_s = MS/(3 − ester-DS), as demonstrated for CB-*graft*-PCL in Sect. 1.2.1. In this chapter, a code CE$_x$-*graft*-PHA$_y$ denotes CE-*graft*-PHA of ester-DS = x and MS = y.

4.2.2 *Thermal Properties: General Transition Scheme*

In the past century, in spite of a large number of precedents of the practical benefit of cellulosic graft copolymers, the relationship between the molecular structure and general thermal transition scheme had not been elucidated thoroughly. This is in a contrastive situation to CEs and cellulose ethers for which excellent systematic works have been conducted, for example, to schematize their thermotropic behaviors [3]. One major reason for the fewer investigations for graft copolymers was that it was rarely accomplished to prepare structurally well-defined graft copolymers. Another point was that interest in grafting had been centered on the surface modification of bulk materials; whereupon it was permitted for the products to be roughly characterized for the application. In 2000s, at length, a substantial progress was made to elucidate the structure–property relationships in a molecular level, using the series of CE-*graft*-PHAs.

Strictly, thermal transition properties are "specimen parameters," which are variable depending on the thermal history of the sample as well as on the measuring method or time-scale. However, the analysis of thermal transition is of significance not only from scientific interest in the generalization of structure–property relationships, but also from a practical point of view as follows: Solids of homo PHAs obtained from lactones often exhibit undesirable mechanical performance due to their lower glass and/or melting transition temperatures (T_g and/or T_m), even if a high degree of polymerization is achieved. In this respect, the copolymerization of these lactones onto cellulosic polymers as sustainer with a high T_g can be a useful application leading to an adequate alteration of the original thermal properties of the PHAs. The systematic analysis data should give an indication of whether the individual drawbacks in thermal properties of the CEs (high T_g) and PHAs (low T_g and/or T_m) can be overcome.

Figure 4.3a illustrates a general scheme of thermal transition behavior for the series of CE-*graft*-PHA as a function of the molecular weight M of the corresponding monomer unit. Actual thermograms in the first and second heating scans are exemplified in Fig. 4.3b, c, respectively, for CA$_{2.15}$-*graft*-PLAs. In Fig. 4.3a, M is defined as a molecular weight per AGU containing ester groups of CE and pendant PHA side-chains. As shown in this figure, T_g decreases monotonically as the unit molecular weight increases from M_0 corresponding to MS = 0 to a critical M_c of MS = N_c. After reaching a minimum at $M = M_c$, T_g gradually increases with M, accompanied by emergence of a crystalline transition. In the following subsections, trials for the formulation are reviewed in detail.

(a) *Glass Transition Temperature*

CE-*graft*-PHA samples with the identical thermal history were examined during the second heating cycle of differential scanning calorimetry (DSC), to clarify the composition dependence of their glass transition behavior [10]. To establish the general scheme of T_g versus composition, two approaches were performed in terms of semi-empirical expressions for (i) homogeneous polymer mixing states, such as random copolymers and miscible polymer blends, and (ii) comb-like polymers.

With regard to the treatment (i), however, it was not suitable to depict the composition dependence of T_g (in K) for the present series of graft copolymers by using the following well-known Fox and Gordon-Taylor equations [10]:

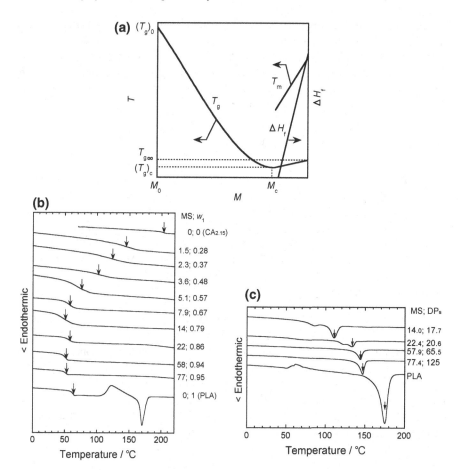

Fig. 4.3   a Schematic representation of the thermal transition behavior of a CE-*graft*-PHA series as a function of the molecular weight M of an esterified AGU which contains an aliphatic polyester side-chain of MS = N, based on DSC thermograms in the **b** 1st and **c** 2nd heating scans illustrated as examples for CA$_{2.15}$-*graft*-PLAs [Reproduced with permission from [10] for **(a)** and [9] for **(b)** and **(c)**]

$$\frac{1}{T_g} = \frac{w_1}{T_{g1}} + \frac{w_2}{T_{g2}} \tag{4.4}$$

$$T_g = \frac{kw_1 T_{g1} + w_2 T_{g2}}{kw_1 + w_2} \tag{4.5}$$

where w_i is the weight fraction of component i; T_{gi} is its glass transition temperature; k is a constant related to the thermal expansion coefficients but practically used as a fitting parameter. Here, the subscripts 1 and 2 denote the PHA and CE components, respectively. As can be seen in both equations, the fixed T_gs are included for the predictions; the poor validity is attributed to the thorough use of a constant T_g for the PHA component.

In general, T_g of a polymer should increase with an increase in its molecular weight. A Fox-Flory relation [11] for T_g of linear polymers with number average molecular weights M_n's may be available to reproduce experimentally observed values; then, the following equation holds:

$$T_{g1}(M_n) = T_{g\infty} - \frac{X}{M_n} \tag{4.6}$$

where $T_{g\infty}$ is an M_n-independent glass transition temperature of the targeted polymer with a sufficiently much higher M_n; X is a polymer-specific constant. When PLA was the PHA component, by substituting $T_{g\infty} = 336$ K for PLA of $M_n = 75.6 \times 10^3$ into Eq. (4.6), the Fox-Flory prediction fitted the experimental T_g versus M_n plot in a range $M_n \geq 1000$, with $X = 3.70 \times 10^4$ calculated by the least-squares method. Concerning T_g of PLA oligomers of $M_n \leq 1000$, however, the following linear fashion rather than Eq. (4.6) was a more practical one to express the M_n dependence.

$$T_{g1}(M_n) = 0.109\, M_n + 189 \tag{4.7}$$

where a limiting value of T_{g1} coincides with 197 K that was T_g obtained for a melt-quenched sample of lactic acid monomer ($M_n = 74$). Figure 4.4a displays a plot of the T_g data for CA$_{2.15}$-*graft*-PLA as a function of PLA weight fraction (w_1), together with some constructions of linear connection between two T_g values, T_{g2} of CA$_{2.15}$ and M_n-dependent T_{g1} of PLA. Open circles in the figure indicate T_g positions at given w_1s, calculated according to the following rule of additivity:

$$T_g = w_1 T_{g1}(M_n) + w_2 T_{g2} \tag{4.8}$$

In the above calculation, $T_{g1} = 197$ K of lactic acid was adopted as far as MS of the graft copolymer was less than unity, which corresponds to a copolymer composition of $w_1 < 0.22$. When the MS exceeded 1.0, Eqs. (4.6) or (4.7) was used for determining the M_n-dependent T_{g1} to substitute into Eq. (4.8). As demonstrated in Fig. 4.4a, the T_gs thus calculated make a curved trajectory on which the T_gs observed for the graft copolymer samples fall closely over the whole range of composition. In particular, the fitting can explain the tendency of gradual elevation of T_g with increasing extent of the PLA-grafting in a range of $w_1 \geq 0.7$.

Another treatment is the one for (ii) comb-like polymers. Insights have been given into effects of the length of normal alkyl side-chains (the number of carbon atom ≤ 18) on T_g of a series of comb-like polymers [12]. For convenience, we can use the structural parameter M (molecular weight per AGU) dealt with in Fig. 4.3a for treating CA-*graft*-PHA series. In analogy with the T_g analysis for the comb-like polymers, by using M and the glass transition temperature $(T_g)_c$ at $M = M_c$

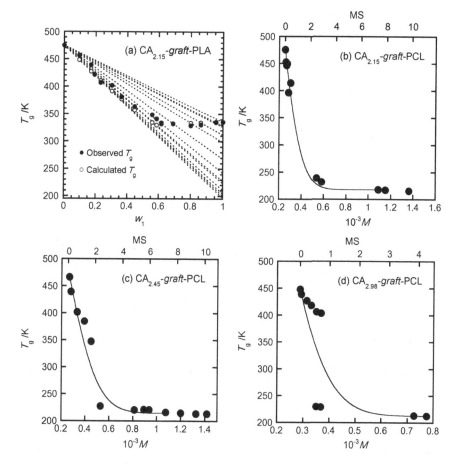

Fig. 4.4 Establishment of the general scheme of T_g versus composition. Two approaches were performed in terms of semi-empirical expressions for **(a)** homogeneous polymer mixing states and **(b–d)** comb-like polymers. (Reproduced with permission from [10]) (see text for discussion)

corresponding to MS = N_c, the copolymer T_g for the $M \leq M_c$ (or $N \leq N_c$) portion of the curve in Fig. 4.3a can be represented in a differential fashion, as follows:

$$\frac{dT_g}{dM} = -K'M\left[T_g - (T_g)_c\right] \tag{4.9}$$

By variable separation and integration of Eq. (4.9) under the conditions that $M = M_0$ and $T_g = (T_g)_0$ for $N = 0$ and $dT_g/dM \to 0$ at $M \to M_c$ and $T_g \to (T_g)_c$, we find a solution:

$$\Delta T_g = \left(\Delta T_g\right)_0 \exp\left[-K\left(M^2 - M_0^2\right)\right] \qquad (4.10)$$

where $K = K'/2$ and

$$\Delta T_g = T_g - \left(T_g\right)_c, \left(\Delta T_g\right)_0 = \left(T_g\right)_0 - \left(T_g\right)_c \qquad (4.11)$$

The coefficient K is actually assessed from Eq. (4.12) as an averaged constant, by using observed T_g versus M data (Fig. 4.4b–d).

$$K = \left\{ -\frac{\ln\left[\Delta T_g / \left(\Delta T_g\right)_0\right]}{\left(M^2 - M_0^2\right)} \right\}_{Ave} \qquad (4.12)$$

Substitution of the averaged constant K and extreme values of $(T_g)_0$, $(T_g)_c$, and M_0 into Eqs. (4.10) and (4.11) enables us to get a calculated smooth curve of T_g as a function of M, as illustrated in Fig. 4.4b in comparison with the corresponding observed T_g versus M plot. We can see a good agreement between the observed T_gs and calculated ones in the plot for $CA_{2.15}$-*graft*-PCL copolymers, and this was also the case for $CA_{1.75}$-*graft*-PCL copolymers [10]. Thus, for these graft series of CAs of acetyl-DS ≈ 2, the decreasing behavior of T_g with increasing side-chain length was represented satisfactorily in terms of the second-order exponential function for the comb-like polymer model.

For additional copolymers using CAs of larger acetyl-DS, however, similar comparison disclosed an evident deviation of observed T_g from the calculated curve, as shown in Fig. 4.4c and d. The T_g data of the $CA_{2.45}$ series (Fig. 4.4c) exhibit a rather positive deviation in a range of $M \leq \sim 2M_0$ and a negative one in $M > \sim 2M_0$, which seems to come from a distinct phase inversion following the change in the dominant polymer component. In Fig. 4.4d, the $CA_{2.98}$ series imparts double T_gs around $M \approx 350$, reflecting a definite phase-separation behavior of the two components constituting the copolymers. It is thus likely that, as the acetyl-DS of the starting CA material becomes higher and higher, the intramolecular density of grafting of the copolymer product should be lowered and, ultimately, the products behave like a "block copolymer." To such a case, the treatment as a comb-like polymer was not applicable any more.

In consequence, we found that the formulation of the composition dependence of T_g proposed for comb-like polymers was adoptable for the graft copolymers of CAs having the original reaction site situated nearby unity in the average number of residual hydroxyls on AGU. The T_g expression based on the comb-like polymer model can serve as an alternative depiction, even though there are some limits in the application involving assessment of multiple data-dependent parameters.

(b) *Melting Transition*

When a polymer is crystallizable, its crystallization progresses with time at temperatures above its T_g. As-prepared samples of the present CA-*graft*-PHA series were, in fact, conditioned by being subjected to vacuum drying (i.e. annealing) at a constant temperature ($>T_g$); hence they usually exhibited a crystalline development of PHA, if the grafted chains were long enough. Such a crystal formation of grafted side-chains had rarely been reported for cellulosic graft copolymers. Melting temperature (T_m) and heat of fusion (ΔH_f) observed by DSC measurements are also "specimen parameters," but it is significant to compare the values as far as the thermal histories of individual samples belonging to the same series are basically identical.

According to a treatment for comb-like polymers [12], the composition dependence of T_m and ΔH_f observed for CA-*graft*-PLA and CA-*graft*-PCL series may be expressed in the following fashions:

$$T_m = T_m^o \left(\frac{M_{side} + \alpha}{M_{side} + \beta} \right) \tag{4.13}$$

$$\Delta H_f = C' + K' M_{side} \tag{4.14}$$

where T_m^o is a melting point of the corresponding PHA homopolyester with a sufficient molecular weight, 458 K for PLA and 342 K for PCL [10]; M_{side} is a number-average molecular weight of polyester chains introduced onto CA, per an originally residual hydroxyl position; α, β, C', and k' are arbitrary constants. M_{side} is determined from $M_{side} = M_a/(3 - \text{acetyl-DS})$ with an available number-average molecular weight M_a of graft chains per AGU, by assuming that the residual hydroxyls of CA are completely replaced by PHA.

In Fig. 4.5, T_m and ΔH_f data are plotted as a function of M_{side} for the graft copolymers CA$_{2.15}$-*graft*-PLA and various CA-*graft*-PCLs. The respective data are found to follow well the functions given by Eqs. (4.13) and (4.14). By extrapolation

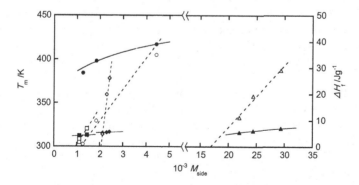

Fig. 4.5 T_m (*solid*) and ΔH_f (*open*) plotted as a function of M_{side} for as-prepared copolymers of CA$_{2.15}$-*graft*-PLA (*black circle, white circle*), CA$_{2.15}$-*graft*-PCL (*black square, white square*), CA$_{2.45}$-*graft*-PCL (*black diamond, white diamond*), and CA$_{2.98}$-*graft*-PCL (*black up-pointing triangle, white up-pointing triangle*). *Solid-line* and *broken-line* curves are drawn according to Eqs. (4.13) and (4.14), respectively (Reproduced with permission from [10])

of ΔH_f versus M_{side} plots to the abscissa, we find a critical M_{side} ($= (M_{side})_c$) required to develop a crystalline phase for each of the crystallizable series examined. Values of $(M_{side})_c$ are estimated roughly as: 1300 for CA$_{2.15}$-*graft*-PLA; 1000 for CA$_{2.15}$-*graft*-PCL; 2000 for CA$_{2.45}$-*graft*-PCL; 17,000 for CA$_{2.98}$-*graft*-PCL. This result clearly demonstrates an increase of the critical M_{side} with lowering density of grafts. In the graft copolymers of acetyl-DS $\geq$ 2.45 showing a lower density of grafts, the adjacent cites for PHA side-chains to anchor onto CA are generally far apart; then, obviously, the side-chain component of shorter length is too disadvantageous to assemble into a crystalline domain, whether the formation takes place intermolecularly or intramolecularly in bulk samples of the graft copolymers. This requires a much larger $(M_{side})_c$ eventually.

4.2.3 *Thermal Treatment Effect on Development of Supramolecular Structures*

As has been reviewed in the above section, the thermal transition property of CE-*graft*-PHAs varies depending seriously upon the copolymer compositions represented by MS. This indicates that the supramolecular structure of the graft copolymers is much alterable by thermal treatment as well as by the copolymer composition. In the past century there had been almost no studies viewing cellulosic graft copolymers from the standpoint of supramolecular structural control by thermal post-treatment; in 2000s, however, there has been a progress in this province by exploring "physical aging" and "crystallization kinetics" for CE-*graft*-PHA [13, 14]. Fortunately, PHAs are preferable subjects for pursuing the thermal treatment effect, because their higher-order structures are susceptible to annealing conditions and the changing kinetics is generally within a moderate time-scale easy to be traced. In a later section, we will describe material functionalization of the CE-*graft*-PHA series in connection with the supramolecular structural development, by exemplifying the enzymatic degradation behavior of the copolymer films subjected to different thermal treatments.

(a) *Physical Aging*

In general, when amorphous materials of polymer are annealed at temperatures below T_g, the excess enthalpy and volume can relax toward those in a more stable glassy state. This is known as enthalpy/volume relaxation or physical aging. An amorphous polymer material aged at a given annealing temperature T_a should recover the relaxed enthalpy at around the T_g in a DSC reheating scan. Figure 4.6a and b display DSC thermograms for two CA$_{2.15}$-*graft*-PLAs (MS = 4.7 and 22, respectively) each aged at 50 °C for different aging periods (t_a). The thermograms demonstrate a steady increase in endotherm with time course. The time evolution of the endotherm ΔH is illustrated in Fig. 4.6c, together with the corresponding data for plain PLA. No crystallization was ascertained to arise in any sample under the

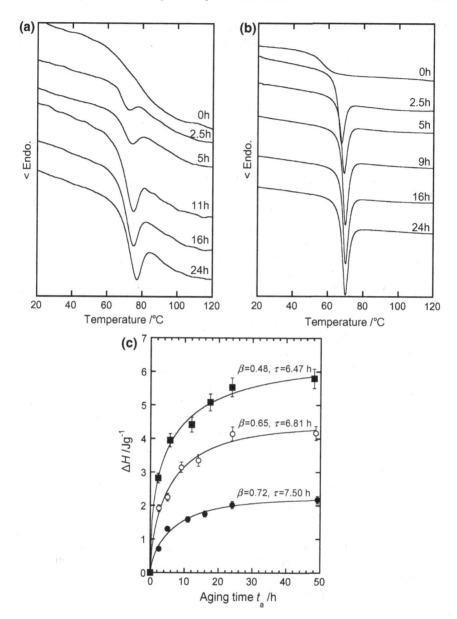

Fig. 4.6 DSC thermograms in the glass transition region for CA$_{2.15}$-*graft*-PLAs of **a** MS = 4.7 and **b** MS = 22 aged for different times at 50 °C, and **c** time evolution of relaxation enthalpy ΔH estimated for plain PLA (*black square*) and CA$_{2.15}$-*graft*-PLAs of MS = 4.7 (*black circle*) and 22 (*white circle*). (Reproduced with permission from [13])

treating condition used. The kinetics of enthalpy relaxation can be described quantitatively by the following Kohlrausch-Williams-Watts (KWW) type of equation with a stretched exponential term [15]:

$$\Delta H = \Delta H_\infty [1 - \Phi(t)] \tag{4.15}$$

with

$$\Phi(t) = \exp[-(t_a/\tau)^\beta] \tag{4.16}$$

where $\Phi(t)$ is the stretched exponential function containing an overall relaxation time τ and a non-exponential parameter β ($0 < \beta \leq 1$). This parameter β is a measure indicating the degree of distribution of the relaxation time; viz., $\beta = 1$ means that there occurs just a single relaxation mode (Debye-type relaxation), while, in contrast, when the width of the distribution is rather broad due to presence of plural relaxation modes in the aging process, the parameter assumes a much smaller value to approach zero.

The result of the KWW analysis indicated that the relaxation process in the graft copolymers follows a rather slower kinetics, compared with that in PLA per se. Of particular interest was the finding of the considerably larger values of β for the copolymers, in distinction to the observation for the PLA providing an index of $\beta = 0.48$ which is just comparable to those (roughly 0.4–0.55) reported for many amorphous polymers [16]. Therefore, it can be deduced that anchoring of PLA graft side-chains onto the semi-rigid $CA_{2.15}$ backbone restricts their mobility and results in a narrower distribution of the enthalpy relaxation mode. The much higher β index for MS = 4.7 may be attributed to an additional structural limit that the PLA side-chains would be too short to interweave mutually.

It is thus expected that the supramolecular arrangement of $CA_{2.15}$-*graft*-PLAs, such as a packing manner of PLA side-chains, can be subtly regulated even in the amorphous state by the aging of moderate term at adequate temperatures.

(b) *Isothermal Crystallization*

As has been shown in the foregoing section, if the PHA grafted side-chains are long enough to be crystallizable, their crystallization progresses with time at temperatures above T_g; the kinetics depends seriously on the copolymer compositions and the degree of undercooling. Plentiful isothermal crystallization data have been gathered to gain insights into the kinetic and morphological aspects of the supramolecular structural development for the CA-*graft*-PLA [13] and CE (CA or CB)-*graft*-PCL series [14].

In the isothermal crystallization experiments, each sample was first heated to attain the isotropic state at a sufficiently high temperature, and successively the sample was quenched to the prescribed isothermal crystallization temperatures (T_{ic}) and held at T_{ic} until the crystallization exotherm was no longer detectable with time course. Figure 4.7a exemplifies DSC exotherms obtained for $CA_{2.15}$-*graft*-PCL$_{9.74}$, being followed with an elapsing time (t) during the crystallization process at the

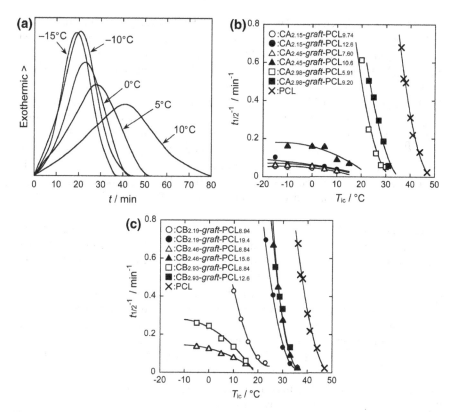

Fig. 4.7 **a** DSC exotherms during isothermal crystallization obtained for CA$_{2.15}$-*graft*-PCL$_{9.74}$ at different T_{ic}s, and plots of $(t_{1/2})^{-1}$ versus T_{ic} for different compositions of **b** CA-*graft*-PCL and **c** CB-*graft*-PCL. (Reproduced with permission from [14])

described T_{ic}. By using the isotherms, the relative crystallinity ($X(t)$) at t can be calculated from integration of the area beneath each exothermic peak. Then, we can coordinate the crystallization kinetics on the basis of the well-known Avrami equation [17]:

$$1 - X(t) = \exp(-Kt^n) \tag{4.17}$$

where K and n are the overall kinetic rate constant and the so-called Avrami exponent, respectively. The rate constant K can also be determined in the following form:

$$K = \frac{\ln 2}{(t_{1/2})^n} \tag{4.18}$$

which includes a half-time of crystallization, $t_{1/2}$, giving $X(t) = 1/2$. Figure 4.7b and c show plots of $(t_{1/2})^{-1}$ versus T_{ic} for different compositions of CA-*graft*-PCL and

CB-*graft*-PCL, respectively. The $(t_{1/2})^{-1}$ reflecting the crystallization rate of the PCL component decreases with an increase in T_{ic} in the temperature range explored. It is also found that the grafting of PCL on the CEs generally causes the crystallization fall down relatively to the situation of the non-grafting. In comparison among the CA-based graft copolymers, the crystallization rates of the two block-like CA$_{2.98}$-*graft*-PCLs are fairly faster than those of the other series of CA$_{2.15}$ and CA$_{2.45}$, in spite of the lower PCL weight content (w_1) in the former series of acetyl-DS = 2.98 (see Table 4.1); this is consistent with the marked domain formability of the CA$_{2.98}$ series. In contrast to this, the $(t_{1/2})^{-1}$ data for the CB-based graft samples (Fig. 4.7c) indicate no systematic dependence of the crystallization rate on the butyryl-DS and side-chain DP$_s$. It can be assumed that the rate declines simply with a decrease in oxycaproyl MS or w_1 in the CB-*graft*-PCL series.

Values of the Avrami exponent n were mostly 1.7–2.5 for the graft copolymers concerned, appreciably small in comparison with n = 3.1–3.4 for plain PCL. The reduction in n can be ascribed to some of the three factors [18]: prevalence of a heterogeneous nucleation; restriction in spatial dimension of the crystal growth; and extremely slow diffusion of the molten chain segments to the crystallization site. In many cases of polymer crystallization where the nucleation takes place preferentially at the surface of a heterogeneous domain present in the molten system, the crystallization of the polymer component would be usually accelerated, often attended by lowering of the growth dimension [19, 20]. For the series of graft copolymers, however, the crystallization rate of the PCL component was generally suppressed, compared with that of ungrafted PCL; besides, a slower growth as "spherulite" was observed for any composition of the CA- and CB-series investigated, as the optical evidence is shown later. From these considerations and the

Table 4.1 Isothermal crystallization kinetic parameters (reproduced with permission from [14])

Sample code	w_1	$T_m^{eq}/^\circ C$	$K_g/10^4$ K^2	$\sigma_e/10^{-5}$ J cm^{-2}
PCL	1	64.8	6.45	0.77
CA$_{2.15}$-*graft*-PCL$_{9.74}$	0.815	49.1	17.9	2.23
CA$_{2.15}$-*graft*-PCL$_{12.6}$	0.851	49.0	17.5	2.19
CA$_{2.45}$-*graft*-PCL$_{7.60}$	0.766	49.9	17.7	2.20
CA$_{2.45}$-*graft*-PCL$_{10.6}$	0.820	50.3	13.9	1.74
CA$_{2.98}$-*graft*-PCL$_{5.91}$	0.701	55.4	11.0	1.35
CA$_{2.98}$-*graft*-PCL$_{9.20}$	0.785	59.0	10.9	1.32
CB$_{2.19}$-*graft*-PCL$_{8.94}$	0.764	56.2	16.2	1.99
CB$_{2.19}$-*graft*-PCL$_{19.4}$	0.875	61.0	12.1	1.46
CB$_{2.46}$-*graft*-PCL$_{8.84}$	0.751	52.1	14.8	1.83
CB$_{2.46}$-*graft*-PCL$_{15.6}$	0.842	62.1	13.5	1.62
CB$_{2.93}$-*graft*-PCL$_{8.84}$	0.733	52.5	16.6	2.05
CB$_{2.93}$-*graft*-PCL$_{12.6}$	0.796	57.1	8.17	1.00

finding of reduced n values, it is natural to assume that the major factor responsible to the diminution of the kinetic parameter is the extremely slower diffusion of PCL chains due to the anchoring onto the semi-rigid cellulosic backbone.

In terms of a kinetic theory of chain-folded polymer crystallization [21, 22], we can estimate the surface free energy of PCL crystals formed in the graft copolymers. The calculation was conducted in the same manner as that described in Chap. 3. The result of calculating the fold-surface free energy σ_e is summarized in Table 4.1. Evidently, any of the graft copolymers provided a σ_e value larger than that obtained for ungrafted PCL. This indicates looser fold surfaces of the PCL lamellae formed in the graft samples. The looseness was conspicuous for lower MS compositions. The σ_e values for the CB-based series were generally smaller than those for the CA-based ones: in the latter, however, the $CA_{2.98}$-*graft*-PCL samples gave an exceptionally small σ_e, probably due to the less prevalence of the diluent effect of the cellulose component in the PCL crystallization. At least for the graft copolymers except the $CA_{2.98}$-*graft*-PCLs, the CE components would be trapped in the interfacial regions between the PCL lamellar crystals, so as to be mixed with the extruded amorphous PCL chains. This process may be smoother in the case of the CB trunk which can be a little more flexible and exhibits much better miscibility with PCL (see Chap. 3), relative to the CA trunk, resulting in the lower roughness of the fold surfaces of the PCL lamellae in the CB-*graft*-PCL series.

In agreement with the isothermal DSC study, polarized optical microscopy (POM) revealed a generally tardy growth of spherulites for the examined CE-*graft*-PHAs (PHAs here include PCL and PLA). During the spherulitic growth, the graft copolymers exhibited no explicit segregation of the non-crystallizable CE component in both the intra- and inter-spherulites under the microscope. The spherulites ultimately impinged on one another and their apparent growth stopped then. Wide-angle X-ray diffractometry confirmed that the crystal structure developed in each copolymer sample was identical with that of PCL or PLA. It is thus natural to assume that the CE fraction was incorporated in inter lamellar regions within the spherulites.

A noteworthy observation was the pattern composed of periodic extinction rings, which appeared only in spherulites of the graft copolymers and never did in plain PHA spherulites, as demonstrated for $CA_{2.15}$-*graft*-PLA samples in Fig. 4.8. The phenomena can be interpreted by a twisted crystal model, in which the adsorption of impurity on crystal boundaries allows the lamellae to twist around an axis of the radial growth [23]. The impurity involves an uncrystallizable polymeric diluent added to the crystallizable host polymer in multicomponent polymer systems. Actually, it was the first time that this type of morphology clearly arouse in PLA spherulites; it had been difficult to form intimately mixed multiple polymer systems containing PLA which is rarely compatibilized by other polymers, in comparison with PCL which is miscible with cellulose nitrate, poly(vinyl chloride), etc. The occurrence of extinction rings for PLA in the graft copolymers virtually supports the universality of that "impurity" hypothesis.

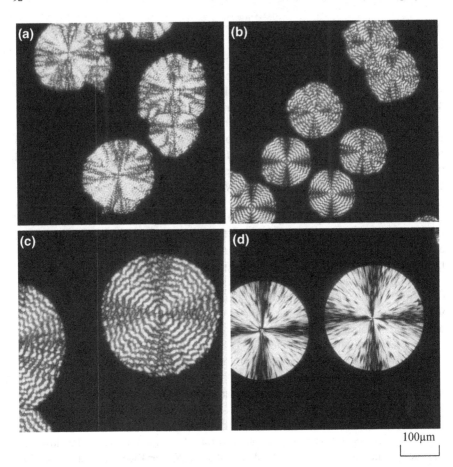

100μm

Fig. 4.8 POM images of typical spherulites observed for CA$_{2.15}$-*graft*-PLAs and plain PLA: **a** MS = 22 at T_{ic} = 107 °C and t = 4000 min; **b** MS = 58 at T_{ic} = 119 °C and t = 365 min; **c** MS = 77 at T_{ic} = 110 °C and t = 300 min; **d** PLA at T_{ic} = 130 °C and t = 30 min, where T_{ic} denotes the isothermal crystallization temperature and t the elapsed time. (Reproduced with permission from [13])

4.2.4 Molecular Dynamics Characterized by Various Relaxation Measurements

The next concern is directed to molecular relaxation analyses by use of different techniques, to deepen understanding of dynamic structures for the CE-*graft*-PHA graft copolymers. As has already been remarked, the graft copolymers are composed of a semi-rigid cellulosic backbone and a flexible aliphatic polymer as side-chains. In such combinations, even though DSC study indicates a single T_g reflecting a considerably homogeneous amorphous mixture, other dynamic measurements can detect separate responses from the two components due to a large difference between their molecular-chain mobility.

For example, we can demonstrate some data of dielectric relaxation spectroscopy (DRS). DRS is a useful method to investigate molecular dynamics of polymers in a widely extended time scale, if the moving sites of the repeating unit and attached side-groups own a permanent dipole moment (see Chap. 1). In the CE-*graft*-PHA copolymers, any of the constituents has a dielectrically active site: –C–O–C– in the cellulose backbone chain and C=O in the acyl groups and PHA side-chains. DRS spectra are generally described in a combination form of the real (ε') and imaginary (ε'') parts of a complex dielectric function. The high sensitivity of the relaxation analysis can readily be seen from a DRS result illustrated in Fig. 4.9, where ε'' data are shown for two samples of CA$_{2.15}$-*graft*-PLA (MS = 22 and 58) [24]. DSC thermograms of these two copolymers of large MS were virtually indistinguishable (Fig. 4.3b). In the ε'' curves in Fig. 4.9, the principal α relaxation corresponding to the PLA glass transition is observed between 320 and 390 K for both, but with a signal essentially split into two peaks. The bimodal signal is very clear for MS = 22 in the lower frequency (f) region of 10^2–10^3 Hz, whereas the bi-mode is much weaker in the other (MS = 58); one peak of the latter sample is degenerated into a small shoulder at temperatures of >350 K. While the peak located in the lower temperature side is associated with the segmental motions of completely amorphous PLA chains, the higher-temperature peak may be attributed to the motion of certain PLA chains dangling from the crystalline domain [25]. The latter mode of relaxation is diminished for the higher MS of 58, because the PLA grafted chains in this copolymer form the crystal domains with higher degrees of stability and crystallinity.

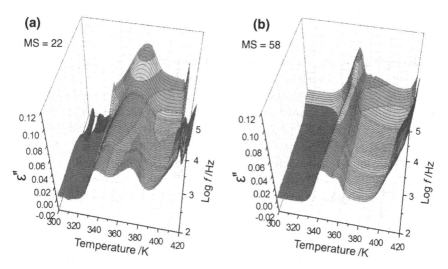

Fig. 4.9 Dielectric loss ε'' curves obtained by DRS measurements for CA$_{2.15}$-*graft*-PLAs: **a** MS = 22 and **b** MS = 58

Kusumi et al. investigated the molecular dynamics and intercomponent mixing state in solid films of the CA-*graft*-PCL and CB-*graft*-PCL series by dynamic mechanical analysis (DMA), DRS, and NMR relaxation measurements [26]. The trunk and graft components of these two series are mutually immiscible and miscible, respectively, in physical blends, as has already been mentioned in the preceding chapter and sections.

The measurement of ^{1}H spin-lattice relaxation time in the rotating frame ($T_{1\rho}^H$) by means of solid-state NMR for specific carbons in a multicomponent polymer system makes it possible to estimate the mixing homogeneity in a scale of ^{1}H spin-diffusion length (2–4 nm) (see Sect. 1.3.2 of Chap. 1). In general, the value of $T_{1\rho}^H$ can be obtained by fitting the decaying carbon-resonance intensity considered to the following single-exponential equation:

$$M(t) = M(0)\exp\left(-t/T_{1\rho}^H\right) \tag{4.19}$$

where $M(t)$ is the intensity observed as a function of the spin-locking time t. Figure 4.10 includes examples of the intensity decay (parts b and c) for selected samples ($DS_{acyl} \approx 2.1$ and MS > 9) of the two CE-based graft series, together with the ^{13}C CP-MAS spectra (part a). Usually, $T_{1\rho}^H$ is simply determinable from the slope in the plot of $\ln[M(t)/M(0)]$ against t. However, some of the logarithmic $M(t)$ data for the PCL-rich copolymers less fitted to a single straight line, because the samples contained a distinct crystalline phase which should give a slower decay of magnetization. In such a case, the normalized $M(t)$ was simulated by a bi-exponential function involving two relaxation times, as follows:

$$M(t)/M(0) = x_f \exp\left(-t/T_{1\rho,\,\text{fast}}^H\right) x_s \exp\left(-t/T_{1\rho,\,\text{slow}}^H\right) \tag{4.20}$$

where $T_{1\rho,\,\text{fast}}^H$ and $T_{1\rho,\,\text{slow}}^H$ represent $T_{1\rho}^H$s of the flexible (faster decay) and rigid (slower decay) components and x_f and x_s are the respective fractions.

Results of the $T_{1\rho}^H$ quantifications for the two CE-*graft*-PCL series are selectively shown in Table 4.2. In both series, as for the compositions of MS < 3 (non-crystalline), all the carbon resonance signals of the CE and PCL components provided only a single $T_{1\rho}^H$; the value decreased with an increase in MS. This decrease reflects that the molecular mobility of the CE-based graft copolymers is enhanced by the escalating introduction of the flexible PCL side-chains as internal plasticizer. In this composition range, both the CA and CB graft series exhibit no remarkable difference in value between the two sets of $T_{1\rho}^H$ data associated with the respective constituting polymers. The phase structure in these copolymers appears to be rather homogeneous in a scale of the path length of the ^{1}H spin-diffusion, i.e., within a few nanometers.

As can be seen in Table 4.2, a critical difference in $T_{1\rho}^H$ allocation manner between the two CE graft series was observed for the compositions of MS > 7; the

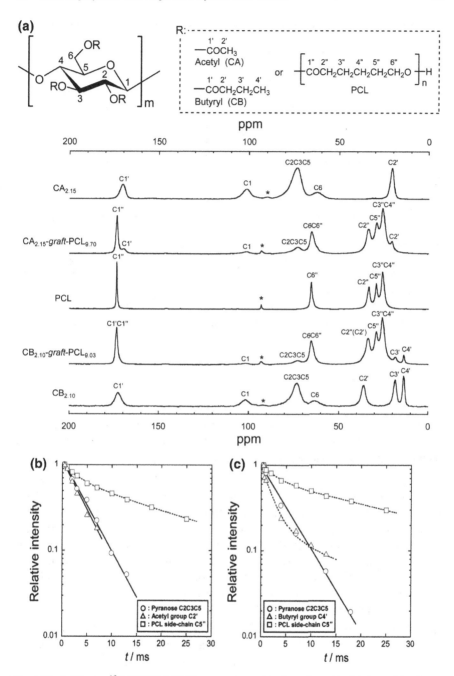

Fig. 4.10 Solid-state ^{13}C CP-MAS NMR spectra (part **a**) and semilogarithmic plots of the decay of ^{13}C resonance intensities as a function of spin-locking time t, for film samples of **b** CA$_{2.15}$-*graft*-PCL$_{9.70}$ and **c** CB$_{2.10}$-*graft*-PCL$_{9.03}$. (Reproduced with permission from [26])

Table 4.2 Data of $T_{1\rho}^{H}$ for CEs, PCL, and selected CE-*graft*-PCLs (reproduced with permission from [26])

Sample code	w_1	$T_{1\rho}^{H}$/ms			
		CA or CB component		PCL component	
		Pyranose C2C3C5	Acetyl C2'or Butyryl C4'	C3"C4"	C5"
CA$_{2.15}$	0	13.7	13.9	–	–
CA$_{2.15}$-*graft*-PCL$_{0.27}$	0.109	12.9	13.3	12.8	12.3
CA$_{2.15}$-*graft*-PCL$_{0.87}$	0.283	8.46	7.87	8.02	8.30
CA$_{2.15}$-*graft*-PCL$_{1.30}$	0.371	6.99	7.11	4.72	3.72
CA$_{2.15}$-*graft*-PCL$_{2.50}$	0.529	2.79	2.74	2.20	1.71
CA$_{2.15}$-*graft*-PCL$_{9.70}$	0.814	4.05	3.79	3.01[a]/24.3[b]	3.63[a]/24.6[b]
CA$_{2.45}$-*graft*-PCL$_{9.30}$	0.800	7.49	8.31	4.10[a]/22.5[b]	4.83[a]/23.3[b]
CB$_{2.10}$	0	9.20	8.64	–	–
CB$_{2.10}$-*graft*-PCL$_{0.16}$	0.0559	7.45	6.73	8.58	6.42
CB$_{2.10}$-*graft*-PCL$_{0.60}$	0.182	6.58	6.23	6.47	5.23
CB$_{2.10}$-*graft*-PCL$_{2.33}$	0.463	3.24	3.11	2.48	2.69
CB$_{2.10}$-*graft*-PCL$_{9.03}$	0.770	3.94	2.02[a]/18.0[b]	3.86[a]/27.3[b]	3.00[a]/31.1[b]
CB$_{2.50}$-*graft*-PCL$_{7.42}$	0.715	4.82	3.20[a]/18.3[b]	3.68[a]/29.4[b]	4.83[a]/31.6[b]
PCL	1	–	–	6.29[a]/60.2[b]	7.17[a]/61.8[b]

[a] $T_{1\rho,\,\text{fast}}^{H}$
[b] $T_{1\rho,\,\text{slow}}^{H}$

plentiful PCL component provides two $T_{1\rho}^{H}$ s due to development of the crystalline phase. In the case where CA was used as the trunk, the magnetization decay of the acetyl C2' resonance yielded a $T_{1\rho}^{H}$ whose value almost coincided with that of another $T_{1\rho}^{H}$ from the pyranose C2/C3/C5 signal, as demonstrated for CA$_{2.15}$-*graft*-PCL$_{9.70}$ in Fig. 4.10b. These values referring to the CA component were situated intermediate between $T_{1\rho,\,\text{fast}}^{H}$ and $T_{1\rho,\,\text{slow}}^{H}$ values referring to the crystallizable PCL component (see Table 4.2). From these observations, it can be deduced that the acetyl group is kept under firm restraint to the cellulose backbone, and the molecular mobility of the unified CA trunk is somewhat restricted by the contiguous PCL crystalline domains. In the case of CB-based copolymers of MS > 7, interestingly, the decay of the butyryl C4' signal was characterized by two different $T_{1\rho}^{H}$ values, while that of the skeletal C2/C3/C5 signal provided a single $T_{1\rho}^{H}$; the specific behavior is exemplified for CB$_{2.10}$-*graft*-PCL$_{9.03}$ in Fig. 4.10c. Plainly, the shorter $T_{1\rho}^{H}$ and longer one quantified for the decayed butyryl C4' resonance can be associated, respectively, with an amorphous phase and with an ordered phase. This observation implies that the butyryl substituent would be considerably free from restraint to the cellulose backbone and, partly, even intrude into the surface region of the PCL lamellar crystals; a similar suggestion has been given in the crystallization kinetic studies [14].

Figure 4.11 shows selected DMA data for CA$_{2.15}$-*graft*-PCLs and CB$_{2.10}$-*graft*-PCLs with different MSs [26]. In this measurement, the dynamic storage modulus E' and loss modulus E'' of their films were collected at 10 Hz as a function of temperature. As can be seen from the data in Fig. 4.11a, the molecular mobility (related to thermoplasticity) of the CE graft copolymers was enhanced, on the whole, by introduction of the flexible PCL side-chains. However, careful observations revealed that the copolymers gave, more or less, a response of "dynamic heterogeneity," despite full assurance of the graft linkage and the good mixing in a scale of a few nanometers. Figure 4.11b and c depict enlarged plots of the E'' data, making clearer various relaxation processes originating from the motions of the relevant structural units in the two CE graft series. Our major concern below is how the principal α_{CE} (α_{CA} or α_{CB}) and α_{PCL} signals, which are associated with the glass transition, i.e., the micro-Brownian motions of the constituting polymers, vary with the composition as a function of MS.

In Fig. 4.11b, we can find the largest dispersion signal (α) located at a temperature position near T_g (calorimetric T_g) for any composition of the CA$_{2.15}$-*graft*-PCLs; nevertheless, another prominent peak appears below or above the T_g, labeled as α'_{PCL} for the CA-rich compositions of MS < 1 and similarly as α'_{CA} for the compositions of MS = 1–2.5. As indicated by dotted lines, the peak positions of α'_{CA} and α'_{PCL} are continuous with the original α_{CA} and α_{PCL} positions, respectively. Thus, in these CA-based copolymer samples under mechanical oscillation, the linked CA and PCL components behave with still mutually different chain-segmental dynamics. Such a dynamic heterogeneity was also found for CB-*graft*-PCL copolymers. However, as seen in Fig. 4.11c, the CB-based copolymers of MS $\approx$ 0.5–0.6 gave a calorimetric T_g just intermediate between the α'_{CB} and α'_{PCL} locations. Additionally, the α'_{CB} signal was not clearly observed at MS > 2, and, reversely, at MS < 0.5, the richer CB component dominated the primary relaxation process of the copolymers. This tendency of assimilation was more prominent in the CB-based series.

DRS analysis made a clear comparison between the two CE-*graft*-PCL series, regarding the cooperativeness in segmental motions of the trunk and graft chains, directly associated with the extent of the dynamic heterogeneity [26]. That is, the cooperative motions were much more conspicuous in the CB-based copolymers than in the CA-based ones. This conclusion was drawn through comparison of the composition dependence of the relaxation time, the activation energy, and the degree of relaxation time distribution for the principal α processes, between the two series. Probably, the butyryl substituent, having a higher structural affinity with a repeating unit of the PCL side-chain, would act as internal compatibilizer to reduce dynamic heterogeneity in the CB-based copolymers; this inference is also supported by the $T_{1\rho}^H$ analysis stated above.

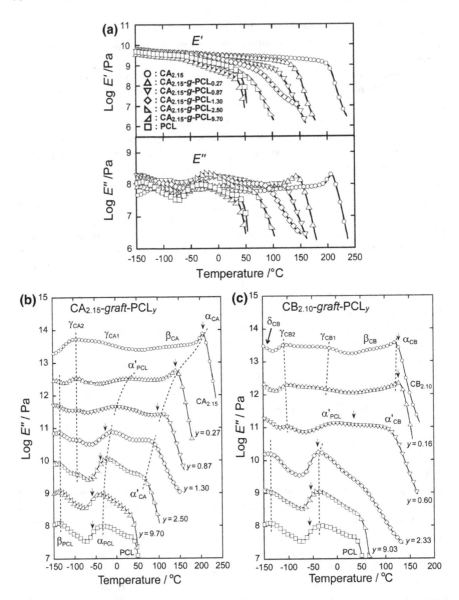

Fig. 4.11 Temperature dependence of the dynamic storage modulus E' and loss modulus E'' for CE-*graft*-PCL copolymers: **a** overall profiles for CA$_{2.15}$-*graft*-PCLs; (**b**) and (**c**) enlarged E'' versus temperature plots for CA$_{2.15}$-*graft*-PCLs and CB$_{2.10}$-*graft*-PCLs, respectively (Reproduced with permission from [26]). In part **b** and **c**, the E'' data are displaced vertically by ~1.0 log unit, relative to the normal position for PCL as reference, and arrows indicate a T_g position estimated by DSC analysis

4.2.5   Tensile Properties and Orientation Characteristics of Films

(a) *Tensile Properties*

Among the series of graft copolymers, combinations of CA (diacetate here) and PLA (or P(CL-*co*-LA)) have been investigated for characterizing tensile properties in the form of film. CA film is generally tough in ambient conditions; tensile strength (σ), % elongation at break (ε), and Young's modulus (E) are ~ 80 MPa, $\sim 20\%$, and ~ 2 GPa, respectively. Tensile properties of PLA film and those of PCL film are in contrast with each other at room temperature; relatively stiff and brittle in PLA ($\sigma = \sim 60$ MPa, $\varepsilon = \sim 20\%$, and $E = \sim 2$ GPa), and flexible and ductile in PCL ($\sigma = \sim 15$ MPa, $\varepsilon \geq 300\%$, and $E = \sim 300$ MPa).

Film sheets of $CA_{2.15}$-*graft*-PLAs were prepared by thermal molding at 180–220 °C followed by quick cooling. The as-prepared films were transparent to the naked eye irrespective of composition. At room temperature, most samples of the $CA_{2.15}$-*graft*-PLA copolymers were so brittle that the tensile behavior could not be fully characterized. The tensile data were therefore collected at selected temperatures of 80–100 °C locating between T_g (~ 200 °C) of $CA_{2.15}$ and that (~ 65 °C) of PLA homopolymer. The grafting was quite effective in giving thermoplasticity to CA. As exemplified in Fig. 4.12a, the data revealed that the ductility of films increased with increasing PLA content and reached its maximum at a certain composition of $w_1 = \sim 0.85$. Above the critical PLA content, the drawability of the films became decreased, due to occurrence of some extent of crystallization of longer PLA side-chains of the copolymers under stretching.

The brittleness observed for the CA-*graft*-PLA series at room temperature can be avoided by copolymerization of lactide and CL for graft chains [27]. For example, film sheets of $CA_{2.45}$-*graft*-P(CL-*co*-LA)s with copolymeric side-chains were ductile and, more or less, endowed with elasticity when the oxycaproyl MS (MS_{CL}) was higher than the lactyl MS (MS_{LA}), whereas there was a fairly steep rise of the tensile strength and modulus with increasing MS_{LA}. The tensile data are plotted in Fig. 4.12b–d. More than an 80-MPa tensile strength and about 70% elongation at break are attained for lactyl-rich and oxycaproyl-rich compositions, respectively. Since the composition dependence of the tensile data well obeys a rule of mixing of copolyester ingredients, the mechanical properties are fairly alterable in compliance with their final applications.

(b) *Orientation Characteristics*

Various processes such as stretching, spinning, and rolling are applied to polymeric materials, to produce films and fibers showing desirable physical properties that vary with direction. This anisotropic behavior is attributed to the preferred orientation of molecular chains. For the thermoplasticized cellulosic graft copolymers, too, a variety of thermal processing methods are applicable and this is expected to

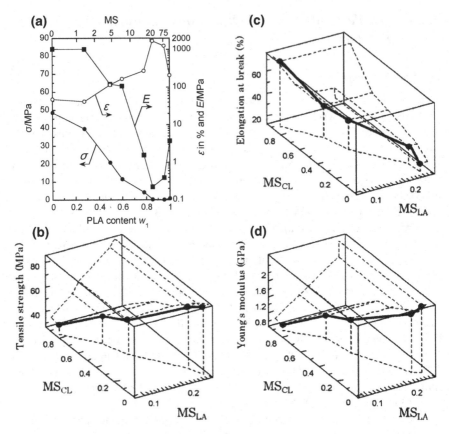

Fig. 4.12 Tensile properties of **a** CA$_{2.15}$-*graft*-PLAs and **b–d** CA$_{2.45}$-*graft*-P(CL-*co*-LA)s. (Reproduced with permission from [9] for **a** and [27] for **b–d**)

invite a specific functionality based on the molecular orientation in the bulk materials.

Unohara et al. [28] examined the molecular orientation and optical anisotropy that were induced by heat stretching CA$_{2.15}$-*graft*-PLA films in connection with the length of their grafted side-chains. For comparison, flexible poly(vinyl acetate-*co*-vinyl alcohol) (PVAVAc) (VA:VAc molar ratio, ~0.36:0.64) was also adopted as a trunk polymer, to offer another series, PVAVAc-*graft*-PLA. The overall behavior of the orientation was estimated in terms of the statistical second ($<\cos^2\omega>$) and fourth ($<\cos^4\omega>$) moments obtained by the fluorescence polarization method (see Sect. 1.4.2 of Chap. 1). All films of both the graft series exhibited a positive orientation function (i.e., $f = ((3<\cos^2\omega> - 1)/2) > 0$) during stretching, which increased with the extent of film deformation. The degree of molecular orientation was higher in the CA$_{2.15}$-graft series with a semi-rigid trunk relatively to the PVAVAc-graft series, but decreased monotonically in both graft series with increasing PLA side-chain content.

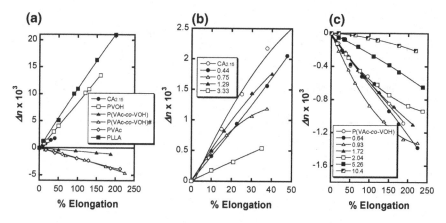

Fig. 4.13 Plots of Δn versus % elongation for film samples of the two series of graft copolymers, CA$_{2.15}$-*graft*-PLA (part **b**) and PVAVAc-*graft*-PLA (part **c**), and of their component polymers and related ones (part **a**). Numerals inserted in (parts **b** and **c**) denote MS values for the graft copolymers. PVAVAc# (in part **a**) indicates an additional copolymer sample with a molar ratio of VA/VAc = 0.164/0.836. (Reproduced with permission from [28])

Concerning the optical anisotropy, CA$_{2.15}$-*graft*-PLA films always exhibit a positive birefringence ($\Delta n > 0$) upon stretching, whereas drawn films of PVAVAc-*graft*-PLA display a negative value, as demonstrated in Fig. 4.13. This contrast in polarity reflects the difference in the intrinsic birefringence between the two trunk polymers. Of particular interest is the discovery of a discontinuous change in Δn with copolymer composition (MS), when compared at a given stage of elongation. This can be explained by assuming different localized orientations of the attached PLA chain-segments; i.e., lactyl units in close proximity to the graft joint are arranged perpendicular to the trunk chain that is most closely aligned to the direction of drawing and make a negative contribution to birefringence, whereas those lactyl units away from the joint are preferentially oriented in the direction of drawing and make a positive contribution to the total birefringence. The situation is illustrated schematically in Fig. 4.14, part (I). Thus, the grafting with elaborate design of the trunk/graft combination ensures not only a wide-ranging control but also a very subtle alteration in the optical property of polymeric bulk materials.

4.2.6   Biodegradation Control

From environmental viewpoints, biodegradability of PHAs was a hot topic from the end of last century to the early 21st century. In 2000 or thereabout, a large amount of data was collected about the biodegradability of PHAs [29, 30]. The cellulosic graft copolymers with PLA or PCL covered in this chapter have also been examined for the enzymatic degradability [31, 32].

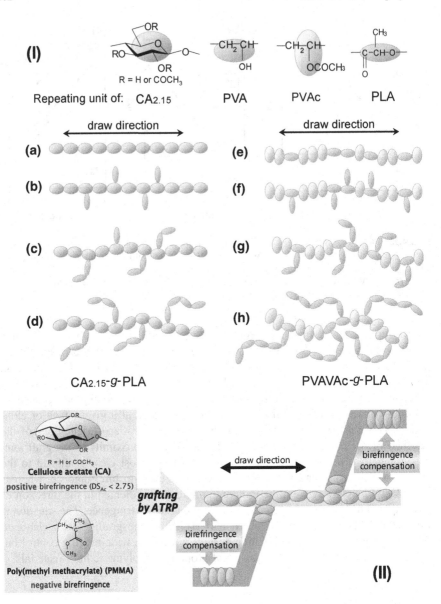

Fig. 4.14 Schematic representations of orientation birefringence for (I) CA$_{2.15}$-*graft*-PLA (**a–d**) and PVAVAc-*graft*-PLA (**e–h**) (see Sect. 4.2.5) and for (II) CA$_{2.15}$-*graft*-PMMA (see Sect. 4.3). Polymeric chains are illustrated as a sequence of the polarizability ellipsoids of the constituent monomeric units. [Reproduced with permission from [28] for (I) and [35] for (II)]

In soil or water, the digestion rate of PLLA (poly(L-lactide)) with high optical purity is generally very slow. In compost, PLLA is practically degraded within half year and meets the ISO standards for biodegradable polymers [33]. With regard to bioabsorbability, a medical test revealed that the half period of decease in strength of PLLA fibers implanted into animals was quite long, ~ 1 year [34]. However, the in vivo digestibility of amorphous poly(D,L-lactide) materials seems to be higher; they were digested within 6 months [34]. These results indicate that the degradability is influenced not only by primary chemical structure but also by the higher-order ones. Contrastively, as for PCL, even though the bulk materials are highly crystalline, the degradation occurs rapidly in both animal bodies and soil.

In the case where multicomponent polymer materials comprising PHAs are used, the difference in degradability between the constituents would make it possible to regulate the overall degradation rate, and even to program the bulk-surface morphology formed during degradation. The control may be accomplished by arrangement of the molecular and supramolecular structures in the bulk materials such as films and fibers. Demonstrative studies were conducted for the CE-*graft*-PHA graft series by the authors.

Enzymatic hydrolysis of $CA_{2.15}$-*graft*-PLLA was carried out for two compositions [MS = 4.7 (L) and 22 (H)] with Proteinase K [31]. The film specimens designated as q-series were solely quenched from the molten state, and other specimens (annealed series) were further heat-treated at temperatures lower or higher than their T_gs. The data of weight loss with elapsing time revealed that the rate of PLLA hydrolysis was depressed by the graft modification itself, and by selecting the lower PLLA content (L) unless the PLLA crystallinity developed. The enzymatic attack on the PLLA side-chains was seriously hindered by the adjacent hydrophobic $CA_{2.15}$ backbone. The heat treatments, which induced the physical aging (for both L and H) or partial crystallization (for H) of the originally amorphous "q-series" specimens, also suppressed the rate and overall degree of enzymatic hydrolysis. These results prove that a temporal control of the enzymatic degradation of $CA_{2.15}$-*graft*-PLLAs is possible not only by varying the molecular compositional factor, but also through the supramolecular rearrangement, i.e., a tighter packing (crystallization) of PLLA grafts and a reduction in free volume (physical aging) in the isothermal treatments.

Regarding the spatial development of the enzymatic hydrolysis, a surface characterization was conducted by atomic force microscopy (AFM) and attenuated total reflection (ATR)-FTIR for degraded copolymer films of the q-series (see Fig. 4.15). The AFM study ensured that the enzymatic hydrolysis caused a transformation into a more undulated surface with a number of protuberances of several-hundred nanometers in height and a few micrometers in width. The ATR-FTIR measurements proved a selective release of lactyl units in the surface region of the hydrolyzed films, and the absorption intensity data was explicable in accordance with the AFM result.

The degraded film specimens often imparted an iridescent color, as illustrated in Fig. 4.15b. The emergence is attributed to the interference of visible lights coming after the diffused reflection between the protuberances formed on the surface of the films, which is virtually interpretable in terms of a thin-layered gradation of refractive index nearby the film surface. These observations of after-effects of the enzymatic hydrolysis for the graft copolymers embody a conception of "spatiotemporally

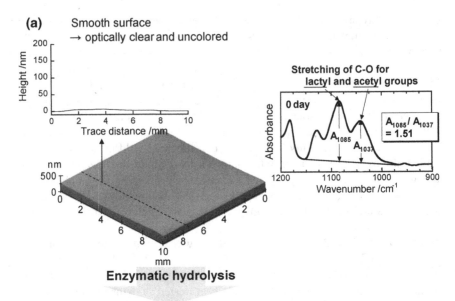

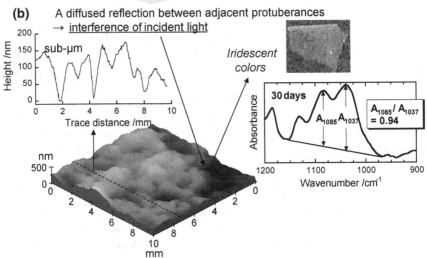

Fig. 4.15 AFM images and ATR-FTIR spectra of a quenched CA$_{2.15}$-*graft*-PLLA$_{22}$ film, obtained (**a**) before and (**b**) after the enzymatic hydrolysis for 30 days. The line profiles of the sections were taken along the dotted lines in the respective AFM images. (Rearranged by using data from [31], with permission)

controlled degradation" for design of novel polymeric materials endowed with multi-functionalities, in which the degradation-rate regulation and the surface modification and ensuing improvement in properties are involved.

4.3   Other Prominences Attained by ATRP: Synthesis and Selected Properties

A refined synthesis technique would yield structurally further defined graft copolymers. For instance, by applying a living polymerization mechanism to prepare cellulosic graft copolymers, we can extremely narrow the distribution of the degree of polymerization DP_s for the grafted side-chains. This allows for us to treat the prepared graft copolymer as cellulose derivative with a fixed size of substituents and therefore to give a delicate functionality to it.

Very recently, Yamanaka et al. synthesized graft copolymers composed of CA (diacetate) and poly(methyl methacrylate) (PMMA) at various ratios by atom-transfer radical polymerization (ATRP) with a CuBr/N,N,N',N'',N''-penta-methyldiethylenetriamine catalytic system [35]. The choice of PMMA was done in consideration of the fact that it is an essential material in optical applications. As shown in Fig. 4.16, these graft copolymers (CA-*graft*-PMMA) were prepared through a three-step process consisting of: (a) the introduction of ATRP initiator (2-bromoisobutyryl groups) onto CA chains to obtain the macroinitiator CAmBBr, (b) ATRP grafting, and (c) the dehalogenation (hydrogenolysis) of the PMMA side-chain terminal. All reactions were conducted in homogeneous solution systems, and the molecular characterization was made in NMR and GPC analyses before and after liberating the PMMA grafts from the cellulosic trunk. These measurements revealed that, even if the molecular weight of the grafts was

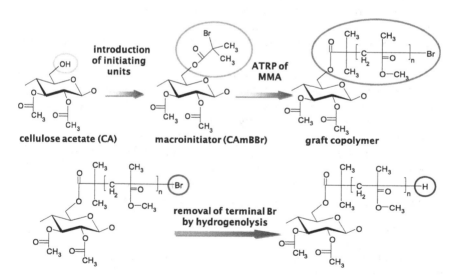

Fig. 4.16 Schematic overview of the synthesis of CA-*graft*-PMMA. (Reproduced with permission from [2])

increased up to $\sim$3000, the polydispersity index remained below 1.2. The graft chains were thus clearly produced via a well-controlled ATRP mechanism.

For a series of CA$_{2.15}$-*graft*-PMMAs, transparent films were made and drawn at temperatures around T_g (125–140 °C), then the drawn samples were examined for evaluation of orientation birefringence Δn. The value of Δn at any stage in the elongation of the copolymer films decreased rapidly with increasing MS (for MMA), eventually leading to a conversion from the positive Δn of pristine CA$_{2.15}$ to a negative value at a composition of >65% PMMA (corresponding to MS = 6.84). Figure 4.14, part (II) presents a schematic illustration of the orientation birefringence for the CA-*graft*-PMMA series. This type of graft copolymer has therefore a potential as high-functional material whose optical anisotropy can be delicately controlled through birefringence compensation between the oriented trunk and graft chains. Especially such a material that can offer zero-birefringence regardless of the deformation would be of great significance in the design of various optical elements (see Sect. 2.2.3 (*b*) in Chap. 2).

4.4 Conclusion and Prospect

As has been reviewed in this chapter, the recently accomplished basic studies on cellulosic graft copolymers made appreciable progress in understanding the structure–property relationships and the microscopic molecular dynamics as well. Particularly, the thermal transition properties of the graft copolymers were well formulated in terms of the molecular and compositional parameters. The accumulation of such inclusive data contributed to developing material functionalities such as controlled degradation and birefringence modulation of molded films, as has been exemplified for several CA-based graft copolymers.

In the synthetic aspect, isolation of pure graft copolymers would pose a perplexing problem for the practical applications. Another progress in precise synthesis such as regioselective substitution of the hydroxyls in AGU of cellulose will further reduce possible structural heterogeneity of the product.

As a matter of course, the concept of grafting in the molecular level is applicable for modification of macroscopic composites of cellulosics. It should also be helpful in creation of advanced materials based on polysaccharide nanostructures such as nanocrystals and nanofibers. Research efforts along this kind of line are expected to more expand the use of cellulose and related polysaccharides.

Acknowledgements The review of this chapter is mainly based on the authors' studies that have been performed in Professor Y. Nishio's laboratory of Kyoto University. The authors express their sincerest gratitude for his excellent guidance and are also grateful to many colleagues in the laboratory for fruitful discussions.

References

1. Nishio Y (2006) Material functionalization of cellulose and related polysaccharides via diverse microcompositions. Adv Polym Sci 205:97–151. doi:10.1007/12_095
2. Teramoto Y (2015) Functional thermoplastic materials from derivatives of cellulose and related structural polysaccharides. Molecules 20:5487–5527. doi:10.3390/molecules20045487
3. Fukuda T, Sugiura M, Takada A, Sato T, Miyamoto T (1991) Characteristics of cellulosic thermotropics. Bull Inst Chem Res Kyoto Univ 69:211–218
4. Buchannan CM, Gardner RM, Komarek RJ (1993) Aerobic biodegradation of cellulose acetate. J Appl Polym Sci 47:1709–1719. doi:10.1002/app.1993.070471001
5. Sakai K, Yamauchi T, Nakasu F, Ohe T (1996) Biodegradation of cellulose acetate by Neisseria sicca. Biosci Biotechnol Biochem 60:1617–1622. doi:10.1271/bbb.60.1617
6. Ajioka M, Enomoto K, Suzuki K, Yamaguchi A (1995) Basic properties of polylactic acid produced by the direct condensation polymerization of lactic acid. Bull Chem Soc Jpn 68:2125–2131. doi:10.1246/bcsj.68.2125
7. Dechy-Cabaret O, Martin-Vaca B, Bourissou D (2004) Controlled ring-opening polymerization of lactide and glycolide. Chem Rev 104:6147–6176. doi:10.1021/cr040002s
8. Moon S, Lee C-W, Taniguchi I, Miyamoto M, Kimura Y (2001) Melt/solid polycondensation of L-lactic acid: an alternative route to poly(L-lactic acid) with high molecular weight. Polymer 42:5059–5062. doi:10.1016/S0032-3861(00)00889-2
9. Teramoto Y, Nishio Y (2003) Cellulose diacetate-*graft*-poly(lactic acid)s: synthesis of wide-ranging compositions and their thermal and mechanical properties. Polymer 44:2701–2709. doi:10.1016/S0032-3861(03)00190-3
10. Teramoto Y, Ama S, Higeshiro T, Nishio Y (2004) Cellulose acetate-*graft*-poly(hydroxyalkanoate)s: synthesis and dependence of the thermal properties on copolymer composition. Macromol Chem Phys 205:1904–1915. doi:10.1002/macp.200400160
11. Fox TG, Flory PJ (1950) Second-order transition temperatures and related properties of polystyrene. I. Influence of molecular weight. J Appl Phys 21:581–591. doi:10.1063/1.1699711
12. Reimschuessel HK (1979) On the glass transition temperature of comblike polymers: effects of side chain length and backbone chain structure. J Polym Sci Polym Chem Ed 17:2447–2457. doi:10.1002/pol.1979.170170817
13. Teramoto Y, Nishio Y (2004) Biodegradable cellulose diacetate-*graft*-poly(L-lactide)s: thermal treatment effect on the development of supramolecular structures. Biomacromolecules 5:397–406. doi:10.1021/bm034452q
14. Kusumi R, Teramoto Y, Nishio Y (2008) Crystallization behavior of poly(ε-caprolactone) grafted onto cellulose alkyl esters: effects of copolymer composition and intercomponent miscibility. Macromol Chem Phys 209:2135–2146. doi:10.1002/macp.200800332
15. Williams G, Watts DC (1970) Non-symmetrical dielectric relaxation behaviour arising from a simple empirical decay function. Trans Faraday Soc 66:80–85. doi:10.1039/TF9706600080
16. Böhmer R, Ngai KL, Angell CA, Plazek DJ (1993) Nonexponential relaxations in strong and fragile glass formers. J Chem Phys 99:4201–4209. doi:10.1063/1.466117
17. Avrami M (1939) Kinetics of phase change. I. General theory. J Chem Phys 7:1103–1112. doi:10.1063/1.1750380
18. Mandelkern I (1964) Crystallization of polymers. McGraw-Hill, New York
19. Takahashi T, Nishio Y, Mizuno H (1987) Crystallization behavior of polybutene-1 in the anisotropic system blended with polypropylene. J Appl Polym Sci 34:2757–2768. doi:10.1002/app.1987.070340811
20. Nishio Y, Hirose N, Takahashi T (1990) Crystallization behavior of poly(ethylene oxide) in its blends with cellulose. Sen'i Gakkaishi 46:441–446. doi:10.2115/fiber.46.10_441
21. Lauritzen JI, Hoffman JD (1973) Extension of theory of growth of chain-folded polymer crystals to large undercoolings. J Appl Phys 44:4340–4352. doi:10.1063/1.1661962

22. Hoffman JD, Frolen LJ, Ross GS, Lauritzen JI (1975) On the growth rate of spherulites and axialites from the melt in polyethylene fractions: regime I and regime II crystallization. J Res Natl Bur Stand Sect A Phys Chem 79A:671–699. doi:10.6028/jres.079A.026

23. Keith HD, Padden FJ, Russell TP (1989) Morphological changes in polyesters and polyamides induced by blending with small concentrations of polymer diluents. Macromolecules 22:666–675. doi:10.1021/ma00192a027

24. Sato M, Kusumi R, Teramoto Y, Nishio Y (2009) Development of supramolecular structures of cellulose acetate-*graft*-poly(L-lactide) in the isothermal crystallization process: X-ray diffraction and dielectric relaxation measurements. 58th Society *of* Polymer Science, Japan Annual Meeting, Kobe, Japan. Polymer Preprints Japan, p 3Pc121

25. Ren J, Adachi K (2003) Dielectric relaxation in blends of amorphous poly(DL-lactic acid) and semicrystalline poly(L-lactic acid). Macromolecules 36:5180–5186. doi:10.1021/ma034420v

26. Kusumi R, Teramoto Y, Nishio Y (2011) Structural characterization of poly(ε-caprolactone)-grafted cellulose acetate and butyrate by solid-state ^{13}C NMR, dynamic mechanical, and dielectric relaxation analyses. Polymer 52:5912–5921. doi:10.1016/j.polymer.2011.10.032

27. Teramoto Y, Yoshioka M, Shiraishi N, Nishio Y (2002) Plasticization of cellulose diacetate by graft copolymerization of ε-caprolactone and lactic acid. J Appl Polym Sci 84:2621–2628. doi:10.1002/app.10430

28. Unohara T, Teramoto Y, Nishio Y (2011) Molecular orientation and optical anisotropy in drawn films of cellulose diacetate-*graft*-PLLA: comparative investigation with poly(vinyl acetate-*co*-vinyl alcohol)-*graft*-PLLA. Cellulose 18:539–553. doi:10.1007/s10570-011-9508-0

29. Li S, McCarthy S (1999) Further investigations on the hydrolytic degradation of poly (DL-lactide). Biomaterials 20:35–44. doi:10.1016/S0142-9612(97)00226-3

30. Tsuji H (2002) Autocatalytic hydrolysis of amorphous-made polylactides: effects of L-lactide content, tacticity, and enantiomeric polymer blending. Polymer 43:1789–1796. doi:10.1016/S0032-3861(01)00752-2

31. Teramoto Y, Nishio Y (2004) Biodegradable cellulose diacetate-*graft*-poly(L-lactide)s: enzymatic hydrolysis behavior and surface morphological characterization. Biomacromolecules 5:407–414. doi:10.1021/bm034453i

32. Kusumi R, Lee S-H, Teramoto Y, Nishio Y (2009) Cellulose ester-*graft*-poly(ε-caprolactone): effects of copolymer composition and intercomponent miscibility on the enzymatic hydrolysis behavior. Biomacromolecules 10:2830–2838. doi:10.1021/bm900666y

33. Funabashi M, Ninomiya F, Kunioka M (2009) Biodegradability evaluation of polymers by ISO 14855-2. Int J Mol Sci 10:3635–3654. doi:10.3390/ijms10083635

34. Miller RA, Brady JM, CD E (1977) Degradation rates of oral resorbable implants (polylactates and polyglycolates): rate modification with changes in PLA/PGA copolymer ratios. J Biomed Mater Res 11:711–719. doi:10.1002/jbm.820110507

35. Yamanaka H, Teramoto Y, Nishio Y (2013) Orientation and birefringence compensation of trunk and graft chains in drawn films of cellulose acetate-*graft*-PMMA synthesized by ATRP. Macromolecules 46:3074–3083. doi:10.1021/ma400155f

Chapter 5
Cellulosic Fiber Produced by Melt Spinning

Yoshitaka Aranishi and Yoshiyuki Nishio

Abstract This chapter focuses on a "melt-spun" cellulosic fiber that was developed through an industry—university cooperative research project. Fiber-formable cellulose is one of the oldest and most familiar apparel materials for humans. However, conventional cellulosic fibers in filamentous forms, such as viscose rayon and acetate rayon, are produced by a solution spinning process that requires conditionally harmful solvents and other reagents. In the present research, the major aim was to design a thermoplastic cellulose derivative and produce filamentous fibers from it using a solvent-free, melt-spinning process. Improved thermal processability of raw cellulose by chemical modification and polymer blending was crucial for successful melt spinning. In relation to the rheological conditioning of the cellulosic composition, we also discuss industrial technical aspects regarding the control of the elongational flow-viscosity and running speed of the filaments in the practical melt-spinning process. Finally, we demonstrate that the melt-spun cellulosic fiber and ensuing textiles exhibit standard properties and distinct functionalities suitable for apparel use. Examples include the facile design of various types of fibers with different cross sections, which are usually not obtained with solution-spun cellulosic fibers.

Keywords Apparel · Cellulose · Cellulose esters · Compatibility · Compositions · Elongation viscosity · Fiber · Melt spinning · Plasticizer · Textile · Thermoplasticization

5.1 Introduction

Cellulose, the largest organic resource on earth, is attracting great interest as a raw material for many applications. Traditionally, commodity use of cellulose is found in fiber or film form. The world's production of fibers (including both bio-based and synthetic fibers) is increasing year after year because of the continuous population increase on the earth. The total demand for fibers in 2014 reached almost 90 million tons per year [1] (Fig. 5.1). This indicates that fiber technology is still important; it

© The Author(s) 2017
Y. Nishio et al., *Blends and Graft Copolymers of Cellulosics*,
Biobased Polymers, DOI 10.1007/978-3-319-55321-4_5

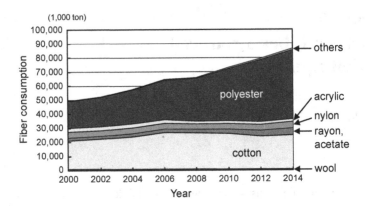

Fig. 5.1 Change of the world's fiber consumption per year since 2000 (graphed using data from [1], with permission)

should be further refined to fulfill the perpetual supply of fibers in the future. In view of the anxiety related to the limited exhaustible fossil resources, it is particularly significant to exploit cellulose for a new type of fiber in an up-to-date fashion.

Established cellulosic fibers form a valuable family that is gaining general acceptance in the apparel field. Cotton is widely used as an apparel fiber, because it excels in hygroscopicity, dyeability, softness, and so on. However, cotton fibers have some limitations because of their innate characteristics concerning dimensions. The fiber length is limited to less than several centimeters, meaning that continuous cotton filaments cannot be formed. The fiber width is mostly in a range 15–25 μm, and the cross section of the fiber is elliptical with a small lumen in the center. Thus the formative characteristic of cotton fibers is largely restricted in the dimension.

There are several well-established methods of spinning to yield cellulosic filaments without a break from the polymer solution. Viscose rayon was invented about 120 years ago. Cellulose is converted into cellulose xanthate in solution using carbon disulfide as a solvent, and then the cellulose xanthate is reduced to cellulose in filament form by forcing the xanthate solution through a spinneret into an acid bath (Fig. 5.2a). A cellulose acetate fiber, acetate rayon, is made by a dry spinning method (without a coagulating medium) using an organic solvent, such as acetone (for cellulose diacetate) or methylene chloride/methanol (for cellulose triacetate). Tencel® (or lyocell) is a cellulosic man-made fiber that is produced by dry-wet spinning of a cellulose solution in N-methylmorpholine N-oxide hydrate (Fig. 5.2b). In any of the cases, inorganic or organic solvents and conditionally other reagents are necessary in the fiber-making process, which is generically called solution spinning.

The problems related to solution spinning are low productivity (i.e., low spinning speed), limitation of the fiber dimension, and the use of the specific solvents or reagents that can be dangerous and harmful to humans and the environment without perfect risk management.

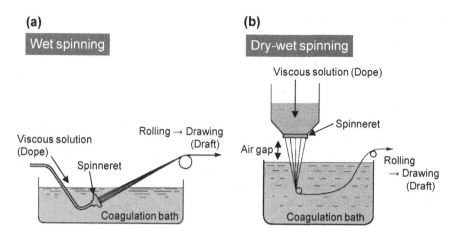

Fig. 5.2 Examples of solution spinning for making cellulose fibers: **a** wet spinning; **b** dry-wet spinning

Melt spinning is a more efficient method for fiber spinning. In application of this method, the processed polymer should have thermoplasticity. Poly(ethylene terephthalate) (PET) and polyamides (e.g., nylon 6) are typical synthetic polymers that can be made into filamentous fibers by melt spinning. The melt spinning method has merits of high productivity (i.e., high spinning speed), larger variations of fiber cross-sections in size and shape, and, most importantly, it is a green process free from harmful solvents. However, the raw materials of the synthetic polyester PET and polyamide nylon 6 are dependent on fossil resources; terephthalic acid, ethylene glycol, and caprolactam are currently made from petroleum. The sustainability of resources is a crucial point in the long-term vision of our life.

In the circumstances mentioned above, we started challenging work to create a new brand of cellulosic fibers [2–4]. The goal of our study was solely to satisfy the two propositions: using a renewable resource, cellulose, as a raw material and producing filamentous fibers by melt spinning without toxic chemicals (see Fig. 5.3).

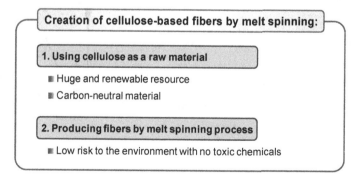

Fig. 5.3 Conceptual slogan of the new challenging work

5.2   Strategy for Melt Spinning Process

Figure 5.4 illustrates a typical melt-spinning process. Pellets of polymer are fed into a vessel of the melt-spinning machine. By simple heating with an internal heater system or a special extruder, the pellets are melted into viscous fluid. The polymer melt is measured up at proper amount with a gear pump and put out continuously through a spinneret apparatus. The effused filaments are cooled in blowing air and taken up with a constant-speed roller, and then wound up with a winder. Thus the melt spinning is a very simple and speedy process for fiber making. As a matter of course, however, the polymer adaptable to this process is of thermoplastic nature, and the melt must be controllable in flow viscosity for the reasonable condition of spinning rate and temperature.

5.2.1   Thermoplasticization of Cellulose

As seen in Fig. 5.5, a PET bottle becomes molten easily by heating, whereas wood-based chopsticks never melt nor soften up on heating, followed by leaving a

Fig. 5.4 Illustration of a typical melt-spinning process

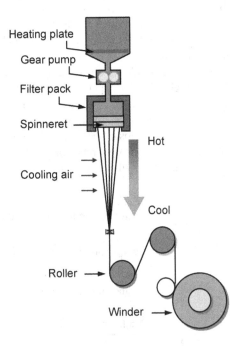

Fig. 5.5 Visual check of thermoplasticity for PET and cellulose

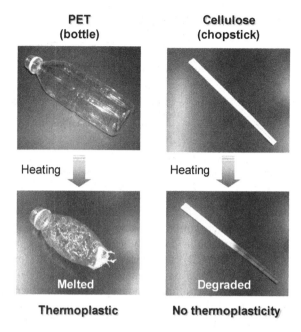

charcoal residue. Thus cellulose does not show thermoplasticity. This is primarily due to a dense network of hydrogen bonds that are formed in and between the molecular chains (Fig. 5.6). No thermotropicity of cellulose means inapplicability of the melt spinning process to this natural polymer itself. To realize the melt spinning using cellulosic material, thermoplasticization of the original cellulose via some treatment is necessary.

Traditional acetate plastics show a certain level of thermoplasticity. In the late 1990s, there were a few patent applications concerning non-woven fabrics made of compositions of plasticized cellulose diacetate [5, 6]. Prior to these, there were research papers of note that reported attempts of melt spinning using hydroxypropyl cellulose [7], phenyl acetoxy cellulose [8], or trimethylsilyl cellulose [9]. In 2000, Gilbert and coworkers again elaborated on the melt spinning mainly using silylether of cellulose [10]. However, any example of the attempts did not reach an industrially usable spinning process, which requires high speed and quite stable conditions for filament production.

From the pioneering researches, at least, it is evident that appropriate chemical modification of the hydroxyl groups of cellulose should be made for our purpose of the thermoplasticization. For example, as shown in Chap. 4, the ring-opening graft copolymerization of cyclic esters onto the cellulose backbone yields a series of thermoplasticized cellulosic compositions. Some of them may be melt-spun without mentioning the fiber performance [11]. Also, as suggested in Chaps. 2 and 3, the acylation of the hydroxyl groups with bulky substituents can provide thermoplasticized cellulosic derivatives. In the simplest acetylation, the product cellulose acetate (CA) may be partially thermoplastic, as can be seen from a photographic

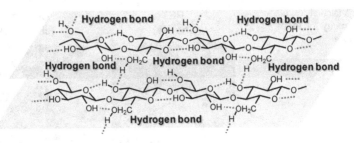

Fig. 5.6 Schematic illustration of a dense network of hydrogen bonds developed in molecular assemblage of natural cellulose

data in Fig. 5.7; the figure compiles visual appearances of several cellulose derivatives each hot-pressed into a film sheet at ~ 220 °C. With substituents of longer chain such as propionyl (for cellulose propionate CP) and butyryl (for cellulose butyrate CB), the thermoplasticity of the resulting cellulose esters more increases and their transparent sheets are readily obtained by hot-press molding (Fig. 5.7, left). A similar tendency resides in etherified celluloses. For instance, ethyl cellulose shows thermoplasticity, while methyl cellulose can hardly become molten at the temperature of 220 °C (Fig. 5.7, right).

Among various cellulose derivatives, cellulose alkyl esters seem to be most easily and economically produced in good control of acylation reaction of raw cellulose. In the familiar acetylation, cellulose pulp in acetic acid is treated with acetic anhydride. After most of the hydroxyl groups are acetylated, the degree of

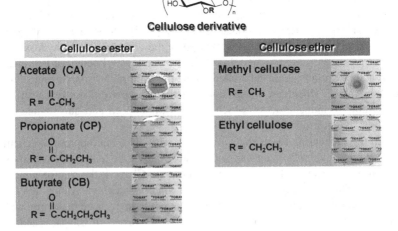

Fig. 5.7 Visual inspection of the thermal processability of cellulose esters (*left side*) and cellulose ethers (*right side*) into a film sheet. Powdered or flaked sample of each cellulose derivative was hot-pressed at ~ 220 °C

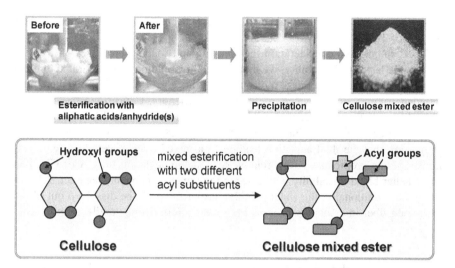

Fig. 5.8 Dual acylation of cellulose into a mixed ester derivative with different two substituents. The *upper* photographic scheme briefly shows a procedure of the esterification of cellulose on a laboratory scale

substitution (DS) is adjusted by hydrolysis treatment of the once acetylated cellulose. Through purification, an acetate lot of cellulose is obtained in powder or flake form. Other alkyl esters including mixed esters (Fig. 5.8) are also producible in a similar or somewhat alterative procedure according to each individual object. The thermoplasticity of the esterified celluloses may be regulated with the adjustable DS and size of the introduced substituents, so as to be suitable for fiber making by melt spinning.

5.2.2   Rheological Conditioning

In the actual industrial process of melt spinning, a controllable range of elongation viscosity must be ensured for the employed thermoplastic cellulose derivative (here, cellulose ester). To meet this demand, it would be convenient and effective to incorporate an additional plasticizer with the cellulose ester.

A traditional low-molecular weight plasticizer for cellulose derivatives is a chemical sort of di-basic acid ester such as 2-ethyl hexyl phthalate, which is, however, suspected to be harmful to the human body (e.g., as an endocrine disturbance). Hazardless glycerin tri-ester, such as triacetin, may be a better plasticizer from an environmental viewpoint. In fact, our earlier trial of fiber manufacturing revealed that the kind of glycerin tri-ester provided a relatively good rheological flow-property as plasticizer for some cellulose esters in the process of melt spinning. However, the plasticizer-containing as-spun fibers were not heat-resistant and exhibited a fateful flaw accompanied by thermal hardening at the stage of their usage as fabrics.

Then, taking into consideration the weaving and soaping processes of fibers as well as the later fabric stability, we thought that it would be more desirable to use a polymer plasticizer of modest molecular weight, e.g., a flexible polyether, poly (ethylene oxide), that is water-soluble and eco-friendly. Concerning the compatibility between the polymer and cellulosic components, it would be attainable by esterifying cellulose with plural substituents (Fig. 5.8). This idea is supported by a suggestion that the dual acylation leading to a mixed ester of cellulose (e.g., cellulose acetate propionate CAP), relative to a single acylation (to CA or CP), can show better blend miscibility with a second polymer component (see Sect. 2.1.1 in Chap. 2). Additionally, the water-soluble plasticizer would be dissolved out of the melt-spun fiber in a soaping process necessary before dyeing/textile arrangement.

5.3   Industrial Example of Success

Based on the preliminary examinations stated above, our target of melt spinning was addressed a compatible system composed of a mixed ester of cellulose and a flexible polyether as viscosity controller. Among several systems of strong possibility satisfying the formula, we first selected a system of CAP/poly(ethylene glycol) (PEG), because of the easier optimization of the chemical composition for melt spinning at reasonable temperatures. In what follows, we briefly outline the practice of melt spinning using the thermoplastic polymer composition comprising CAP, PEG, and a trace amount of antioxidant, as an example of successful industrialization. Toray Industries, Inc. named and registered a trademark *Foresse*® for the first successful melt-spun cellulosic fiber.

5.3.1   Manufacturing Process

Figure 5.9 shows a flow chart of the industrial process for manufacturing the cellulosic fiber *Foresse*® and ensuing textile. The CAP used is produced by esterification of the original raw cellulose using acetic acid/anhydride and propionic acid (or anhydride). By compounding the CAP (flake), PEG, and a particle of additive at a prescribed proportion of the optimum in a heated twin-screw extruder, transparent pellets of the polymer composition can be obtained [see photographs in Fig. 5.9 (right)]. The melt spinning process for fiber making is similar to that applied to synthetic polymers such as PET and polyamides, which is shown in Fig. 5.4 and includes three major steps: (1) the pellets supplied in succession are re-melted and correctly measured through the gear pump; (2) the molten fluid then enters into the filter pack and put out from the spinneret; (3) after quick cooling, the fiber spun in multi-filaments is finally taken up by the winder through a sequence of rollers. As

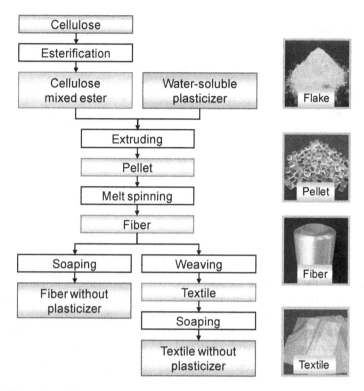

Fig. 5.9 Flow chart of the industrial process for manufacturing the cellulosic fiber *Foresse®* and textiles using the fiber

far as the compatible polymer composition was designed to show an adequate range of the melt viscosity and that of the elongation viscosity, the continuous melt spinning into filamentous fibers was successfully conducted.

After fiber manufacturing by melt spinning, the cellulose-based fibers should be arranged into textiles by weaving or knitting (see Fig. 5.9). In the soaping process indispensable for dyeing, the water-soluble PEG component was removable from the textile. Of course, this elimination was possible for the as-spun fibers before weaving. Thus, it was ascertained that the fibers and textiles without plasticizer are also obtained in a safety manner easy on the environment.

5.3.2   Flow Properties

As regards the present application of melt spinning, there is no prominent difference in the manufacturing procedure between the thermoplastic cellulosic fiber and conventional synthetic fibers. In reality, however, flow properties of the polymers quite differ from each other. In Fig. 5.10, elongational flow-viscosity data are

Fig. 5.10 Elongational flow-viscosity of the cellulose-based filamentous effusion measured right after spinning, in comparison with the corresponding data for melt-spun PET as reference (see text for discussion)

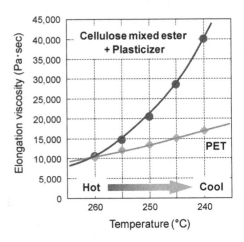

compared between the two polymer species, one being the cellulose-based one and the other PET; the data were measured for their single filamentous effusions right after spinning. The abscissa denotes the periphery temperature changing according as the filament is put out from the spinneret. In the case of PET, the elongation viscosity increases gradually with a decrease in the temperature. In contrast, the thermoplasticized cellulosic composition shows a sharp rise of the elongation viscosity. This difference may be eventually attributed to the difference in molecular chain flexibility between the two polymer species; the cellulose backbone is comparatively stiffer.

Figure 5.11 makes a comparison of a so-called fiber speed (i.e., running speed of filament) in the melt spin-line between the above two cases; the data are plotted as a function of the distance from the spinneret. In this illustrative experiment, the steady-state spinning speed (i.e., rate of fiber winding) was set at 2000 m/min. In

Fig. 5.11 Fiber speed in the melt spin-line for the cellulose-based composition and a reference PET, plotted as a function of the distance from the spinneret for filament effusion. This comparison is made at a steady-state spinning rate of 2000 m/min

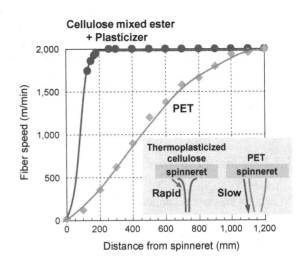

the case of PET, the fiber speed increases gradually with the slow increase in the viscosity (Fig. 5.10). From this situation, the running PET filament will be completely solidified at a position of ~100 cm below the spinneret. In the other case using the cellulosic composition, the fiber speed increases very rapidly. At a short distance of ~20 cm from the spinneret, the fiber speed already reaches 2000 m/min. Thus the spin-line profile of the running filament concerned was specific and different from those of PET's and common nylon's.

As instructed by the above data, careful and stable control of the sharp spin-line is very important in the melt spinning of the cellulosic composition. Recent advance in hardware of the spinning technique largely contributed to clear the issue without difficulty, however. Actually, continuous winding of the melt-spun fiber *Foresse*® is easily accomplished at 2000 m/min (= 120 km/h), more than twice the speed of the solution spinning established for other cellulosic fibers.

5.4   Fiber and Textile Properties

Our next concern was "whether or not the successful melt-spun cellulosic fiber would be practicable especially in the apparel field", that is, "whether the fiber and textiles would exhibit standard properties and some distinct functionality suitable for apparel use". The answer is yes, as demonstrated below.

5.4.1   General Properties for Clothing

While the thermoplasticized cellulose derivative is not pure cellulose but based on its ester, fabrics made of the fiber show excellent properties in hygroscopicity, dyeability (adsorption) and coloring, electrostatic protection, etc. This is chiefly because of a certain amount of hydroxyl groups remaining on the cellulosic backbone, and, possibly, partly due to the introduced carbonyls of proton-accepting nature. Major properties of the fiber and textile are listed in Table 5.1 in comparison with PET fiber as reference.

In respect of mechanical property, the melt-spun cellulosic fiber is not categorized as high-performance fiber of the present day, but yet it shows moderate tensile behavior. As exemplified in Fig. 5.12, the cellulosic fiber exhibits a higher tensile strength, compared with solution-spun acetate rayon fibers of diacetate (D) and triacetate (T). Differing from PET or common polyolefins (e.g., polyethylene), cellulose esters generally do not show development of high-crystalline structure under a rheological process from melt. This was also the case for the present cellulosic composition. Instead, the fiber showed a relatively high degree of molecular orientation when obtained in certain high-speed conditions of the melt spinning. Presumably this orientation development is responsible for the better tensile strength of the cellulosic fiber relative to those of the acetate fibers from

Table 5.1 Properties of the cellulosic fiber *Foresse*® (after soaping) and its woven textile adaptable for various clothes, together with data of PET fiber as reference

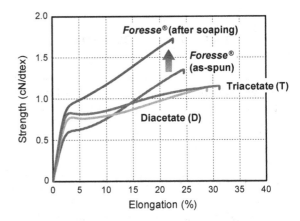

Melt-spun fiber	Refractive index	Modulus (cN/dtex)	Standard moisture (%)	Electrostatic potential (kV)
PET	1.58	80~120	0.4	9
Foresse®	1.48	30~50	4.0	2~3

Textile Properties: Lustrous, Vivid, Soft, Hydrophilic

Apparel Use: Outer, Formal dress, Blouse, Sports wear, Inner, Lining

Fig. 5.12 Tensile mechanical behavior of the cellulosic fiber *Foresse*®, compared with data for cellulose acetate fibers produced from solution (*note* 1 cN/dtex ≈ 130 MPa)

solution. In Fig. 5.12, it is also interesting to notice that the cellulosic fiber offers an appreciably higher strength after soaping treatment. Evidently, the elimination of flexible polymer plasticizer enhances the tenacity of the as-spun fiber.

Figure 5.13 serves photographs of fabric examples (a) knits and (b) taffeta cloths made of the cellulosic fiber *Foresse*®. As can be perceived from the picture in Fig. 5.13a, the knits afford a soft touch, probably proceeding from the modest modulus and good flexuous nature of the fiber. In Fig. 5.13b, we can see vivid and lustrous colors of the taffeta cloths. As shown in Table 5.1, the refractive index of the cellulosic fiber is 1.48, which is considerably lower than that (1.58) of PET. Therefore, colors come out better from the former medium of lower refractive index, due to less reflection loss of visible light. With regard to moisture uptake (see Table 5.1), while PET fiber is quite poor in hygroscopic nature, the cellulosic fiber shows a high-standard value of moisture gain. Accordingly, this fiber can offer adequate hydrophilic textiles endowed with a good anti-static property; a supporting data (electrostatic potential) is also included in the table.

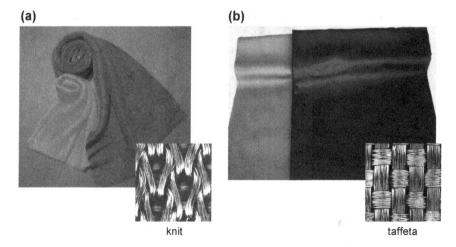

Fig. 5.13 Fabric examples, **a** knits and **b** taffeta cloths, made of the cellulosic fiber *Foresse*®

5.4.2 Shape Variation and Functional Diversity

A great merit of adoption of the melt spinning is that we are able to fix a variety of cross sections distinctive in shape for the produced fibers. Traditional cellulosic filaments, such as viscose rayon and acetate rayon, are obtained by solution spinning. In these productions, the filamentous fiber put out from the spinneret will make its shape while being separated from the solvent of the initial solution. Plainly, it is not so easy to control the fiber cross-section in the presence of the solvent during the spinning process. On the contrary, in the melt spinning that requires no solvent, the molten filament effused from the spinneret is solidified with simple blowing of cooling air. The sectional shape of the solid filamentous fiber is substantially similar to that of the spinneret hole, except for special cases. Therefore, the cross-section of the melt-spun fibers may be easily controllable by designing the shape of the spinneret holes. This held good for the cellulosic fiber *Foresse*® manufactured successfully by melt spinning.

Figure 5.14 demonstrates various types of the melt-spun cellulosic fiber designed in different shapes of the cross section. Besides the normal round type (photo **a**), fibers of tri-lobed type (photo **b**), X-type (photo **c**), S-type (photo **d**), and hollow type (photo **e**) are collected. These variations can alter the standard textile properties (based on the normal type **a**), for instance, in luster, water uptake, or lightness of weight, which will contribute to diversify functionalities of the final apparel products.

Another example of hollow fiber is illustrated in Fig. 5.15 (upper); the porosity of this fiber is 40%, and its apparent specific gravity (sp. gr.) is 0.8 lower than that of water. Therefore, as shown in the figure (bottom), a bundle of the fibers can float on or beneath the water surface for some time. With solution-spun rayon, it is usually hard to produce such an "ultra-light" fiber.

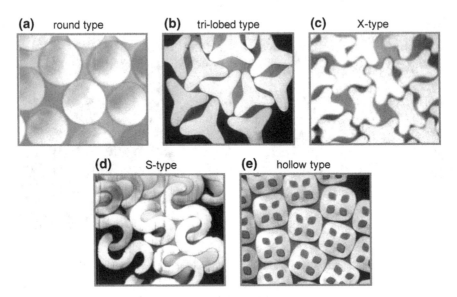

Fig. 5.14 Various types of the melt-spun cellulosic fiber designed in different shapes of the cross section: **a** normal round type; **b** tri-lobed type; **c** X-type; **d** S-type; **e** hollow type. The diameter of single filament of each fiber is 10–20 μm

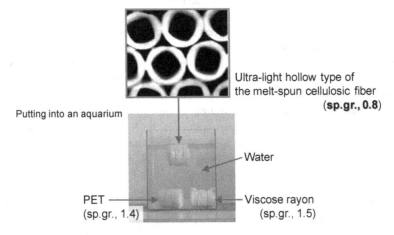

Fig. 5.15 High functional hollow-type of cellulosic fiber, achieved by melt spinning (*upper*). The ultra-lightness of the hollow fiber is demonstrated by floating behavior of the fiber bundle in an aquarium (*bottom*). Additional fibrous bundles of PET and viscose rayon as reference are made of their respective normal type of fibers without hollow

Figure 5.16 (upper) shows a conjugated sea-islands-type fiber; the diameter of an apparently single filament is ∼20 μm in this example. This fiber was obtained by spinning of a molten mixture of polymers that showed a phase separation of sea-islands type. As the island component, the thermoplastic cellulosic composition

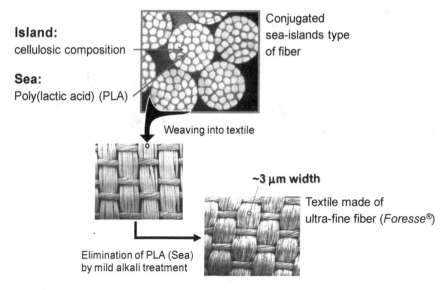

Fig. 5.16 Conjugated sea-islands type of fiber obtained by melt spinning (*upper*) and a textile woven of the fiber (*middle*). Formation of ultra-fine cellulosic fiber (*Foresse®*) is also demonstrated (*bottom*) (see text for discussion)

was used. As the sea component, poly(lactic acid) (PLA) was used because the degradable PLA would be removed from the conjugated fiber by mild alkali treatment. Acting on the expectation, the melt-spun fiber was made into a woven textile (Fig. 5.16, middle) and then the textile whole was alkali-treated. Obviously, the sea component PLA was readily eliminated and the island component remained as fine filaments (Fig. 5.16, bottom). In this demonstration, each single diameter of the newly appearing filaments is ~ 3 μm, smaller than those of cotton fibers by one order magnitude. This result indicates that "ultra-fine" fibers can also be achieved by melt spinning of the thermoplastic cellulosic composition.

5.5 Concluding Remarks

With the aim of creating a new cellulosic fiber that is producible by melt spinning, we studied thermoplasticization of cellulose and rheological conditioning of the designed cellulosic composition. Cellulose per se is thermally intractable because of the dense hydrogen-bonding network in molecular assemblage. However, appropriate chemical modification of the side groups on the cellulose backbone can bring a good thermoplasticity for processable material and also offer much opportunity for compatibly mixing with a second polymer. In the industrially successful melt spinning exemplified above, the thermoplasticization of cellulose by dual acylation (i.e., mixed esterification) and the regulation of elongational flow-viscosity of the

Fig. 5.17 Exhibition of textile products using the cellulosic fiber *Foresse*®

cellulose derivative by compatibilizing with a flexible polyether were crucial. In spite of the sharp thinning profile of filaments running in the unsteady state, a current basic process of melt spinning was readily applicable to the thermoplasticized cellulosic composition. In view of some aspects concerned with the environment and resources, this success must be of great significance.

As regards the practicability of the successful melt-spun cellulosic fiber, we also demonstrated that the fiber and textiles show standard properties and distinct functional characteristics suitable for apparel use; some of the characteristics are hardly realized with solution-spun cellulosic fibers. For example, it is possible to design the cellulosic fiber with various cross sections (e.g., of tri-lobed type, conjugated sea-islands type, and hollow type) with great facility by the adoption of melt spinning. Undoubtedly, this contributes to enhancement of the functional diversity of the apparel products. In actual fact, as shown in Fig. 5.17, the brand-new fiber has recently invited much opportunity of textile exhibition and gained acceptance in apparel industry as a readily processable, comfortable, and sustainable fabric element.

Acknowledgements A part of this research was financially supported by the New Energy and Industrial Technology Development Organization (NEDO) of Japan.

References

1. Japan Chemical Fibers Association (ed) (2015) Sen'i handbook 2016. pp 166–174
2. Nishio Y, Aranishi Y (2006) Thermoplasticized cellulose fibers, part 2, chap. 1. In: Hongu T (ed) Super-biomimetic fiber technologies. NTS Pub, Tokyo
3. Aranishi Y, Ichikawa T, Yamada H, Nishio Y (2007) Melt spinning of thermoplasticized cellulose. In: 2nd International Cellulose Conference (ICC 2007), Tokyo, Japan. Book of abstracts, p 156
4. Aranishi Y, Nishio Y (2008) Fabrication of cellulosic fibers by melt spinning, part of developments, chap. 19. In: Isogai A (ed) Advanced technologies of cellulose utilization. CMC Pub, Tokyo
5. Maurer G, Rustemeyer P, Teufel E (Rhodia Acetow GmbH) (1997) Melt-blown non woven fabric, process for producing same and the uses thereof. WO 9733026

6. Matsubayashi Y, Tsujimoto N (Oji Paper Co, Ltd) (1997) Biodegradable cellulose acetate fiber and its production method. Japan Patent, H09-291414

7. Shimamura K, White JL, Fellers JF (1981) Hydroxypropylcellulose, a thermotropic liquid crystal: characteristics and structure development in continuous extrusion and melt spinning. J Appl Polym Sci 26:2165–2180. doi:10.1002/app.1981.070260705

8. Pawlowski WP, Gilbert RD, Fornes RE, Purrington ST (1988) The liquid-crystalline properties of selected cellulose derivatives. J Polym Sci Polym Phys Ed 26:1101–1110. doi:10.1002/polb.1988.090260514

9. Cooper GK, Sandberg KR, Hinck JF (1981) Trimethylsilyl cellulose as precursor to regenerated cellulose fiber. J Appl Polym Sci 26:3827–3836. doi:10.1002/app.1981.070261129

10. Gilbert RD, Venditti RA, Zhang C, Koelling KW (2000) Melt spinning of thermotropic cellulose derivatives. J Appl Polym Sci 77:418–423. doi:10.1002/(SICI)1097-4628 (20000711)77:2<418:AID-APP19>3.0.CO;2-I

11. Aranishi Y, Nishio Y (2006) Thermoplasticization of cellulose and its application to the melt spinning. Cellulose Commun 13:70–74

Printed in the United States
By Bookmasters